ECONOMIC ISSUES, PROBLEMS AND PERSPECTIVES

ECONOMICS OF INNOVATION, INCENTIVES AND UNCERTAINTY

Economic Issues, Problems and Perspectives

Additional books in this series can be found on Nova's website under the Series tab.

Additional e-books in this series can be found on Nova's website under the e-book tab.

Global Economic Studies

Additional books in this series can be found on Nova's website under the Series tab.

Additional e-books in this series can be found on Nova's website under the e-book tab.

ECONOMIC ISSUES, PROBLEMS AND PERSPECTIVES

ECONOMICS OF INNOVATION, INCENTIVES AND UNCERTAINTY

PATRICK E. SIMMONS
AND
SAMUEL T. JORDAN
EDITORS

nova
publishers
New York

For permission to use material from this book please contact us:
Telephone 631-231-7269; Fax 631-231-8175
Web Site: http://www.novapublishers.com

Library of Congress Cataloging-in-Publication Data

Economics of innovation, incentives and uncertainty / editors, Patrick E. Simmons and Samuel T. Jordan.
p. cm.
Includes bibliographical references and index.
ISBN 978-1-62257-251-9 (hbk.)
1. Industries. 2. Industrial policy. 3. Research, Industrial. 4. Technological innovations--Economic aspects. I. Simmons, Patrick E. II. Jordan, Samuel T.
HD2326.E239 2012
338'.064--dc23

2012021144

Published by Nova Science Publishers, Inc. † New York

CONTENTS

PREFACE

The authors have gathered research from across the globe in the study of the economics of innovation, incentives and uncertainty. Topics include incentives for SME's in the European Union; technology innovation and economic success in Japanese manufacturing firms; nonpoint pollution policy in the presence of uncertainty; the use of DEA-DA for corporate value assessment; the impact of system of trade preferences (non-reciprocal) on development outcomes in beneficiary economies; the innovation and export activity causal link; uncertainty in the labor and financial markets and the sluggish U.S. macroeconomic performance; the implications of strategic delegation in a mixed oligopoly model; monetary uncertainty and the demand for money in the U.S.; and bonus/promotion schemes as incentives.

Chapter 1- The aim of this chapter is to illustrate, in largely systematic manner, the financial incentives currently adopted in some Countries of the European Union (EU) for small and medium-sized enterprises (SMEs), and to place them within the framework of the contextual policy adopted at community level for this important as well as problematic component of the European productive system.

The initial part of this chapter consists of an analysis of the Community's legislation on so-called 'state aid', while the final part pays particular attention to the more recent innovative tools, aimed at encouraging the sources of risk capital to finance SMEs. In this way an attempt is made to highlight the advantages, disadvantages and profiles of coherence of the EU strategy.

Chapter 2- This study examines governance relationship between a board structure and corporate performance in Japan. It is easily imagined that a support from the top management is essential for both technical innovation and economic success in corporate activities. In other words, the board structure may determine technical innovation and economic success as the most influential management component. In this research, the firms sampled for this study consist of 280 Japanese manufacturing firms. The first section of Tokyo Stock Exchange Market lists all the firms examined in this study. The observed period is from 2003 to 2006. The empirical results partly support the effectiveness of the ongoing efforts on a board structure reform in Japanese firms. This study identifies the three empirical findings. First, a size reduction in the board does not increase the corporate performance of Japanese manufacturing firms. Second, the introduction of independent directors enhances the corporate performance. Finally, the high independency of directors further improves the corporate performance.

Chapter 3- Water quality throughout the world is declining in response to diffuse pollution from agricultural production. It is difficult to highlight cost-effective means to assess alternative policy instruments for its regulation because of the presence of much uncertainty regarding annual pollutant flows, agent heterogeneity, climate and soil characteristics, and future price movements. Bioeconomic modelling represents a flexible means to assess the suitability of different policy instruments in the presence of such uncertainty. A general framework that considers alternative on-farm mitigation technologies in different land uses in agricultural catchments is introduced for the analysis of such problems. The model is applied to the Lake Taupo catchment in New Zealand to identify the abatement costs associated with emissions standards of different intensities and the relative value of market-based and uniform policy instruments. Emission reductions beneath 30 per cent are found to incur minimal abatement costs when appropriate mitigation technologies are adopted. This reflects the low abatement cost present in the simulated dairy sector and the cost-effective mitigation practices available to achieve reasonable leaching reductions on sheep and beef farms, which constitute more than 90 per cent of the catchment. Moreover, this finding is robust to broad changes in key model parameters. Thus, land use heterogeneity is highlighted as an important determinant of whether such policies are appropriate for the regulation of agricultural activity to improve water quality.

Chapter 4- This study discusses a use of DEA-DA (Data Envelopment Analysis-Discriminant Analysis) to assess the corporate value of firms. It is widely known among finance researchers that Tobin's q is a ratio to measure the corporate value. A contribution of this study can be found in a linkage between DEA-DA (as a methodology) and Tobin's q (as a theoretical basis). As an application of the combined use between DEA-DA and Tobin's q, this study investigates the influence of R&D (Research & Development) expenditure on the performance of Japanese manufacturing industries. This research identifies that the R&D expenditure is important for the Japanese manufacturing industries to be competitive in a global market. However, the importance of R&D expenditure depends upon the type of an industry. Empirical findings identified in this study are useful to other industries (not only manufacturing business) and other nations (not only Japan).

Chapter 5- This chapter argues that current approaches to studying the impact of trade system preferences (non-reciprocal) on development outcomes in beneficiary economies have not been quite comprehensive in coverage. Especially with respect to the objective of promoting industrial development, most studies rely only on static impact issue of export facilitation to pass judgment on success of program implementation. The dynamic aspect of the analysis has largely been ignored. Combining insights from the literature on trade preferences and globalization, the present chapter proposes a three-dimensional impact analysis model that incorporates elements of both static and dynamic issues. These issues concern the roles the program can play in facilitating exports, enhancing competitiveness of producers and fostering emergence of linkage related integrated manufacturing. It is hoped that a vintage framework of this sort will help guide future endeavors towards more comprehensive analytical engagements.

Chapter 6- In this work, the authors review the literature on the link between innovation and exports. In particular, they try to study the causal direction of the relation between both variables: it is whether innovation improves the export activity of the firm, as suggested by the self-selection theory, or otherwise, whether it is the greater firm export activity what enhances the innovative performance of the firm, as predicted according to the learning-by-exporting arguments. The authors focus on a number of recent studies that, dealing with the endogeneity concern, show stronger support for the self-selection theory on innovation, rather than for the reverse link, stated by the learning-by-exporting hypothesis. This conclusion matches up with the most common finding of the export literature when considering the firm productivity.

Chapter 7- Uncertainty in labor and financial markets has received a great deal of blame for the United States' recent sluggish macroeconomic performance. As a result, it can be expected that future studies will focus on the role of this risk in determining a wide variety of macroeconomic variables. How is uncertainty in economics measured? In this chapter the authors review different measures of uncertainty that are widely applied in the literature, focusing on those that are based on variance, prediction error, and GARCH modeling. For demonstrative purposes, the authors generate these different measures of uncertainty for the monthly dollar-pound real exchange rate and compare their statistical properties. In general, all of the measures capture the same periods of high volatility, but their statistical properties vary widely.

Chapter 8- This paper examines the implications of strategic delegation in a mixed oligopoly model when the government imposes an environmental tax to mitigate pollution damage. Managers of both public and private firms are compensated by a linear combination of profit and sales revenue. The effects of strategic delegation by comparing the equilibria with and without strategic delegation are explored. It shows that when strategic delegation is applied, the government sets a lower environmental tax. It is opposite to Bárcena-Ruiz and Garzón's (2002) results in pure oligopoly with managerial delegation, but concurs with the finding obtained in Bárcena-Ruiz and Garzón (2006) and Wang and Wang (2009) in mixed market settings without managerial delegation.

Chapter 9- The current prolonged recession in the U.S. has necessitated the Fed to engage in expansionary monetary policy, to a degree of which is above and beyond most people's expectation. Such quantitative easing has eroded both consumer and business confidence due to the fear of inflation. The Nobel Laureate Milton Friedman (1984) argued that the increased volatility of money growth raises the degree of perceived uncertainty and therefore, increases the demand for money. This is known as Friedman's "volatility hypothesis".

This chapter revisits the "volatility hypothesis" in the U.S. by estimating the demand for the M2 monetary aggregate. A standard money demand that includes a measure of monetary uncertainty, in addition to income and interest rate, is estimated using recent advances in econometrics literature and annual data over 1971-2010. Friedman's hypothesis is supported, implying that the current prolonged recession in the U.S. could partly be due to the increased holding of cash because of monetary uncertainty.

Chapter 10- Why are bonus/promotion schemes so widely used in reality? Are they effective in alleviating incentive problems? For the standard agency model, this paper proposes an alternative solution to the classical solution in Holmström (1979). The advantages of the author's solution are that (a) it is a simple solution; (b) it is not based on the troublesome first-order approach; and (c) it is the first best. The disadvantage of the author's solution is that it imposes a boundary condition on the support of the distribution function. However, this boundary condition does have sensible economic interpretations and the optimal contract resembles the widely observed bonus/promotion scheme in reality. Hence, the author's solution shows the potential of a bonus/promotion scheme in resolving incentive problems.

In: Economics of Innovation, Incentives and Uncertainty
Editors: P. Simmons and S. T. Jordan
ISBN: 978-1-62257-251-9

Chapter 1

INCENTIVES FOR SMES IN THE EUROPEAN UNION, WITH PARTICULAR REFERENCE TO ITALY

Aurelio Bruzzo, Jacopo Mattei and Stefania Zanzi[*]
Department of Economics and Management, University of Ferrara, Italy

ABSTRACT

The aim of this chapter is to illustrate, in largely systematic manner, the financial incentives currently adopted in some Countries of the European Union (EU) for small and medium-sized enterprises (SMEs), and to place them within the framework of the contextual policy adopted at community level for this important as well as problematic component of the European productive system.

The initial part of this chapter consists of an analysis of the Community's legislation on so-called 'state aid', while the final part pays particular attention to the more recent innovative tools, aimed at encouraging the sources of risk capital to finance SMEs. In this way an attempt is made to highlight the advantages, disadvantages and profiles of coherence of the EU strategy.

Keywords: Financial incentives, SMEs, EU regulations, 'state aid', Italian policies

1. INTRODUCTION

1.1. Types of Incentives and Their Definitions

According to economic theory, incentives are instruments with which the state seeks to induce a particular form of behaviour by various actors (consumers or enterprises) considered to be rational and, therefore, also sensitive to economic stimuli.

[*] Although the present chapter is the result of a joint work by the authors, the responsibility of Jacopo Mattei is for sections 4 and 5.6, of Stefania Zanzi for sections from 2.1 to 2.6, of Aurelio Bruzzo for the remainders.

This chapter will deal with the incentives furnished – within a particular scheme – by public administrations to productive enterprises in order to foster the pursuit of a series of economic policy goals, such as an allocation of resources better than that determined by the free market, or promotion of a Country's products with respect to those of its competitors (Del Monte 1994).

The range of economic incentives that can be granted by the state is rather broad, and it comprises diverse types. Firstly, incentives can be grouped according to their aims: to stimulate production investments by enterprises; to increase employment, research and development; and, more recently, also to promote energy saving, environmental protection, labour mobility, and so on.

Besides the aims of incentives, it is then necessary to distinguish between general ones, which concern enterprises belonging to all production sectors or all geographical areas, and specific incentives, which instead concern small groups of industries and geographical areas. Therefore there may be incentives for sectors, for certain industries and categories of firms, or for particular regions.

Incentives can be further distinguished between those targeted on particular factors (capital and labour, for instance) and those targeted on production. The former can then be divided into two categories according to whether they involve outlays by the government (financial incentives) or the foregoing by the government of future payments by enterprises (fiscal incentives). Belonging in the former category are financial subsidies (or capital grants), interest subsidies, premiums per labour unit employed; while belonging in the latter are tax credits, accelerated depreciation, exemption of social security contributions, and particular tariff allowances concerning the use of various forms of transport, electricity consumption, etc.

Besides the various types of fiscal and financial incentives just outlined, since the 1970s numerous European Countries have adopted policies intended to furnish public support for small and medium-sized enterprises (SMEs) through the supply of real incentives: that is, services connected with the establishment and management of businesses. These policies have been prompted by recognition that the obstacles to the growth of enterprises are not only of a financial kind; they very often also derive from the fact that firms, especially in late-developing areas, do not have access to real services which are satisfactory from a qualitative-quantitative point of view – for financial reasons or because of a lack of supply.

Real incentives, in their turn, comprise a wide range of services: (a) organizational and managerial assistance; (b) technological assistance (for instance, through the acquisition and transfer of technologies and processing systems); (c) commercial assistance (marketing and information about markets, etc.); (d) financial and administrative assistance. The models adopted to implement policies on real incentives are of two types patterned on the experiences of the UK and Germany.

In order to prevent 'commercial wars' provoked by increasingly favourable incentives systems, international treaties define the rules with which EU Member States must comply. For example, certain Articles in the EEC Treaty set out Community rules in regard to "state aid". The principle, still in force, on which this regulation was based is that any aid granted must respect Community rules in regard to free competition. Aside from the fact that the concept of 'aid' now introduced is broader than the previous one of 'incentive', in that it encompasses all measures intended to favour certain enterprises within a competitive market, the Community principle of the incompatibility of state aid is not absolute; indeed, the EEC Treaty defined certain categories of aid compatible with the common market.[1]

Another remark preliminary to the present chapter concerns the long-standing debate among economists about the convenience of tax bonuses compared with capital incentives.

On the one hand, there is a group of economists (for example those belonging to the "*Associazione per lo Sviluppo del Mezzogiorno*" (SVIMEZ) in Italy) who have sought to contrast the EU Commission's intent to exclude the granting of inducements and other financial incentives to areas internal to the Member States of the EU and instead relate them to the Country as a whole. To this end, these economists have argued that historical experience, including that of Italy's southern regions, does not seem to bear out the contention that public incentive policies are substantially ineffective or even harmful. The issue, instead, is how to increase the coherence of those policies with their original goals[2].

On the other hand, a larger group of economists have reached the conclusion that the policy of capital incentives for businesses conducted in Italy during the 1990s produced largely disappointing results, since subsidized enterprises achieved a turnover per employee which was less than that of non-subsidized enterprises. Moreover, the effects of the subsidies appeared to cause distortions because – in that they did not affect labour productivity – they produced distortions for large firms operating in sectors with greater bargaining power. Instead, tax benefits would have yielded decidedly more satisfactory results, positively influencing the demand for skilled labour, as well as the introduction of new products and new production processes by the beneficiary enterprises.

1.2. Aims and Contents of the Chapter

It is evident from the discussion in the previous section that the picture is still contradictory from various points of view. Consequently, a study on incentives such as the present one, which takes stock of the situation from both the normative and empirical points of view, is entirely opportune.

In fact, the aim of this chapter is to develop a reconstruction, as exhaustive as possible, of the current state of affairs in regard to the main economic policy measures of a financial nature, referable to the category of incentives, more recently adopted for SMEs in Italy in light of the corresponding measures assumed at the level of both the EU and the other member-Countries.

The focus is on Italy for various reasons: firstly, because the Italian economic system is marked by a large number of SMEs which indubitably perform a decisive role in regard to the main macroeconomic variables (production, investments, exports, etc.); secondly, because of

[1] These categories were: aid to promote economic development in regions where the standard of living was abnormally low, or where there was severe underemployment; aid intended to promote the execution of an important European project or to remedy a serious disturbance of the economy; aid intended to facilitate the development of certain activities or certain economic regions, provided that such aid did not alter trading conditions in a manner contrary to the common interest; other categories of aid as defined by decision of the Council on a proposal by the Commission.

[2] It should also be borne in mind that more recent surveys (e.g. Cardosi, 2010) have shown, amongst other things, a correlation between investments made by micro and small enterprises and the facilities granted. In the case of investments by enterprises of this kind, which together with medium-sized enterprises form the core of the Italian production system, the facilitation instrument also conforms to the 'additionality' requirement stipulated by EU rules.

the particular attention traditionally paid by the public institutions, both national and regional, to this important component of the overall production system – so much so that now available is a large and complex body of legislation and relative measures; thirdly, because of the recent attempt to adjust the framework by introducing and using further innovative instruments with which to tackle the difficulties caused by the severe crisis ongoing at international level.

Consequently, the study is structured into three main parts:

- the first part (section. 2) considers 'state aid' as regulated by the EU legislation currently in force and as recently used by the main European Countries, among which obviously Italy;
- the second part (section 3) sets out the EU's enterprise policy, with particular reference to SMEs;
- the third part (sections 4 and 5) illustrates the recent SME policy implemented in Italy, also through application of the Community document entitled *The Small Business Act*.

The method of analysis with which each of the various parts of the study are conducted will be the one – economic, legal and managerial – which appears best suited to the topic addressed. Hence the study will be interdisciplinary in nature.

The bibliographical references will therefore also be heterogeneous, in that they will consist not only of scientific studies in the business, the economic, and the legal fields but also of official documents pertaining both to the European Union and Member States.

2. 'State Aid' According to EU Legislation and in the Recent Experience of Some European Countries

2.1. 'State Aid' According to Recent EC Regulations

It is advisable to begin with brief discussion on the term 'incentives' as it is used in what follows. It has been rightly pointed out that "the word 'incentives' does not have an unequivocal meaning in the Italian legislation, and that "Italian law does not contain definitions of the fillip activity. The term denotes a series of actions whose denominations are multiple and various: contributions, subsidies, grants, facilitated financing. By 'incentives' is meant all public financial outlays to subjects which undertake economic activities complying the public will or which consent to engage in a certain behaviour wanted by the public authority".[3]

---

[3] Cit. Tortorelli (1992), p. 211, and also Di Plinio (1998), p. 598, for whom "the term 'incentivization' and similar expressions (incentives system, aid, support for businesses) denote legal techniques instrumental to the achievement of objectives generally definable as public. Such techniques are intended to induce particular subjects (generally private enterprises) to adopt economic behaviour conducive to attainment of the results desired. Accordingly, incentives are often defined in unitary manner as *planning instruments*. In reality, the instruments in question have been historically highly diversified, with measures often applicable solely in the specific sector of reference, and with techniques difficult to allocate among scientifically verifiable categories. Moreover, the legislative terminology is often inconsistent and imprecise".

We shall therefore use the expression 'incentives measures' in the sense of "means to attribute advantages to the private entrepreneurs that adopt behaviour and/or achieve results compliant with the public will".[4]

The doctrine[5] concerned with intervention by the public powers[6] as "expressed through the 'suasive' and consensual attribution to private actors of advantages of various kinds for the purposes of general economic policy"[7] has addressed diverse issues: for example, the legal definition of the various type of public aid,[8] the features of the state financier[9] and its capacity to manage the economy,[10] regional laws for economic development[11] and their compatibility with international[12] and EU regulations restricting the sovereignty of states in regard to their power to apply incentives measures.[13]

It is obvious that this is not the place for re-examination of these issues, in regard to which the reader is referred to the ample literature already produced. In what follows, we shall concentrate on the last aspect mentioned, describing the features of current[14] law on 'state aid' with particular regard to SMEs.

For the purposes of this study, it should be emphasised that 'incentives' – like 'aid' to businesses – constitute a matter of prime importance for European Union law, and they are therefore made subject to the regulations on state aid[15] with a view to creating and maintaining markets open to free competition.

---

[4] Cit. Cavallo and Di Plinio (1983), pp. 419 and 422, where one reads that " legal science predominantly defines the technical notion of incentives as the attribution of an advantage to an enterprise with the purpose of fulfilling a public interest ".

[5] The bibliography on public incentives for businesses is enormous. See for present purposes Amorth (1931), Carabba (1970), Pericu (1971), Serrani (1971), Giannini (1977b), Cossu (1984), Croci and Pericu (1990), Castaldi (1998) and Manzella (2003).

[6] On the expression "public powers" see Giannini (1986).

[7] Cit. Giusti (1997), p. 393.

[8] See e.g. Croci and Pericu (1990), p. 244, for whom in Italian law, "the financial aid system is configured as a complex and heterogeneous set of measures with profoundly different structural characteristics but united by the identity of the economic function performed. In French jurisprudence see, most recently, Colin (2011), p. 276.

[9] On the expression "state financier" see Serrani (1971).

[10] On this see Giannini (1977a), p. 228 and Giusti (1997), p. 393, who writes " besides governance of the economy by 'authoritative means', increasingly common is influence on the economy expressed through the 'suasive' and consensual attribution to private actors of advantages of various kinds for the purposes of general economic policy. These actions to support or foster economic activities take the form of a myriad of instruments and operational solutions of great diversity and difficulty of classification. Because they almost always arise from improvised choices of political expediency, it is difficult to discern a systematic set of public aid measures in Italian law, so that it seems more correct to speak of a series of aid policies implemented in different ways and with different instruments."

[11] The term «economic development» may recall the topic of competition. Economic development must be accompanied by the development of competition, to which reference will be made below. See the distinction between "development" and "protection" according to Article 117 II c., letter e) of the Italian Constitution. For our purposes, see Pappalardo (1990), p. 113, Trovato (1993), p. 191, Ammassari and Marzorati (1998), and Tonetti (2003), p. 597.

[12] For completeness, mention should be made of the international regulation of subsidies. On this see in particular Di Comite (2009), p. 7, where he points out that "the current discipline divides among diverse legal regimes according to specific categories of subsidies and their effects on international trade".

[13] On incentive measures see in particular, Cocozza (2007), p. 226, which develops issues already discussed in legal theory, especially in the 1960s. See Loiodice (1968), p. 787, and Pace (1983), p. 2341.

[14] For a study on the evolution over time of legislation on state aid see Tosato (2011), p. 3.

[15] For a synthesis of legislation on state aid see European Commission (2008).

2.2. 'State Aid' in the Treaty on the Functioning of the European Union

In order to frame[16] European Community regulations on state aid correctly, reference must necessarily be made to the Treaty of Rome signed on 25 March 1957 (which established the European Economic Community). Among the instruments proposed to create the common (now single) market, in fact, the Treaty cited the restriction of state aid to enterprises.

The rationale at the base of this regulation is evident: the rules on aid granted by Member States are intended to ensure that such aid does not unduly benefit certain enterprises or products with respect to competitors, thereby distorting competition in the market.

It should be borne in mind that when the Treaty of Lisbon came into effect (1 December 2009), the Treaty establishing the European Community (TEC) was renamed[17] Treaty on the Functioning of the European Union (TFEU), and that the Articles, sections, chapters, titles and parts of the Treaty on the European Union and of the Treaty on its functioning were renumbered. Here, therefore, we will refer to the consolidated version of the Treaty on the European Union and of the Treaty on the Functioning of the European Union.[18] The rules on state aid are set out in title VII (common rules on competition, taxation and approximation of laws), chapter 1 (rules on competition), section 2 (aid granted by states). The main provisions on state aid are contained in Articles 107-108-109 of the TFEU (respectively ex arts. 87, 88 and 89 of the TEC).

Before examining its content, we would stress that, according to Article 3 (paragraph b) of the TFEU, the Union has exclusive competence in various areas, including definition of the competition rules necessary for the functioning of the internal market, and that the competition rules comprise not only those relative to aid granted by states (the subject of the discussion here), but also the rules applicable to enterprises contained in section 1 (Articles 101 to 106) and relative to agreements, abuse of dominant position, and concerted practices.

2.3. Continued: Prohibited and Permitted Types of 'Aid'

That said, we may ask what 'aid' is prohibited and what is instead permitted? We reiterate that the notion of 'state aid' is difficult to define because it is not 'enshrined' in any law. In fact, the Treaty does not regulate state aid but merely prohibits any 'measure'[19] which matches the features specified in Article 107 (ex Article 87 of the TEC), section 1 of the Treaty, viz. "any aid granted by a Member State or through state resources in any form whatsoever which distorts or threatens to distort competition by favouring certain undertakings or the production of certain goods".

According to Article 107, I c, "Save as otherwise provided in the Treaties, any aid granted by a Member State or through state resources in any form whatsoever which distorts or threatens to distort competition by favouring certain undertakings or the production of

[16] See, among many others, De Carli (1995), p. 132, Picozza (2001), Carullo (2005), p. 126 and Cassese (2012), p. 56.

[17] Pursuant to Article 2, paragraph 1 of the Treaty of Lisbon.

[18] The consolidated versions of the Treaty on the European Union and the Treaty on the Functioning of the European Union were published in the Official Gazette no. C 83 on 30 March 2010.

[19] This is not the place for close examination of the notion of 'law provision', although useful for understanding the use of the term 'measure'. See Huber (1963) and Mortati (1968).

certain goods shall, in so far as it affects trade between Member States, be incompatible with the internal market".

The first paragraph of the Treaty Article thus furnishes the notion of aid and enables us to delimit the provision's range of application. It stipulates the incompatibility of state aid with the internal market, but as yet makes no definitive statement on the range of the prohibitions that this entails. In fact, the principle of incompatibility is not absolute; rather, it allows derogations for aid granted for specific purposes: incompatibility is, therefore, subject to a twofold series of derogations.

The second paragraph of the Article identifies a first group of derogations that we may regard as compatible *de jure* with the internal market.[20] The following are compatible with the internal market: aid of a social nature; aid to remedy the damage caused by natural disasters or other exceptional events; and aid granted to the economy of certain regions of the Federal Republic of Germany in order to compensate for the economic disadvantages caused by the Country's division at the end of the Second World War.

The third paragraph defines the aid that can be considered compatible with the internal market "by virtue of a decision adopted by the Commission (or exceptionally by the Council) on the basis of assessments of a broadly discretionary kind".[21] In fact, considered compatible with the internal market are the following: aid to promote the economic development of areas where the standard of living is abnormally low or where there is serious underemployment; aid to promote the execution of an important project of common European interest or to remedy a serious disturbance in the economy of a Member State; aid to facilitate the development of certain economic activities or of certain economic areas (provided that it does not adversely affect trading conditions in a manner contrary to the common interest); aid to promote culture and heritage conservation where such aid does not affect trading conditions and competition in the Union to an extent that is contrary to the common interest; and other categories of aid as may be specified by decision of the Council on a proposal from the Commission.

2.4. Continued: Controls

Under Article 108, the Commission is the supervisory body. What is the 'procedural mechanism' envisaged by the Treaty? Firstly, the Treaty draws a distinction between new aid and existing aid. In the latter case, Article 108, I and II c., foresees three types of intervention.[22] Firstly, the Commission, together with Member States, maintains constant supervision, proposing to states any appropriate measures with which to limit aid. Secondly, the Commission takes the initiative by inviting the parties concerned to submit comments, and if it then finds that the aid is incompatible with the internal market it orders the state to abolish that aid, or alter it in the manner that it has decided. Thirdly, if the state concerned does not comply with the Commission's decision within the established term, the Commission, or any other interested party, may apply directly to the Court of Justice of the European Union for adjudication.

---

[20] On this see Roberti (1997), p. 11.
[21] Cit. Roberti (1997), p. 12.
[22] On this see Cassese (2012), p. 37.

In regard to new types of aid (plans to grant new aid or to alter already-existing forms), III c. of Article 108 provides for a system of preventive control. This latter consists in a double obligation on states: firstly, Member States must inform the Commission of any plan to introduce or alter state aid (*ex-ante* notification requirement); secondly, Member States cannot implement the measures that they envisage before they have been evaluated and authorized by the Commission ("suspension principle"). If states grant aid without the Commission's approval, such aid is automatically considered "illegal". The following Article 109 of the Treaty "acts as the legal basis for the issue of derived acts of law; it empowers the Council to adopt regulations for the application"[23] of Articles 107 and 108.

The provisions of the Treaty are then completed with the issue of various regulations. Principal among the latter are the following:

- Council Regulation (EC) no. 994/98 of 7 May 1998 on the application of Articles 92 and 93 of the Treaty establishing the European Community to certain categories of horizontal state aid.[24]
- Regulation (EC) no. 659/1999 of 22 March 1999 laying down rules on application of Article 93 of the EC treaty.[25]
- Commission Regulation (EC) no. 68/2001 of 12 January 2001 on the application of Articles 87 and 88 of the EC Treaty to training aid.[26]
- Commission Regulation (EC) no. 70/2001 of 12 January 2001 on the application of Articles 87 and 88 of the EC Treaty to state aid to small and medium-sized enterprises.[27]
- Commission Regulation (EC) no. 794/2004 of 21 April 2004 implementing Council Regulation (EC) No 659/1999 laying down rules for the application of Article 93 of the EC Treaty;[28] and Commission Regulation (EC) no. 1125/2009 of 23 November 2009 amending Regulation (EC) No 794/2004 implementing Council Regulation (EC) no. 659/1999 laying down detailed rules for the application of Article 93 of the EC Treaty, as regards Part III.2, Part III.3 and Part III.7 of its Annex I.[29]
- Commission Regulation (EC) no. 1628/2006 of 24 October 2006 on the application of Articles 87 and 88 of the Treaty.[30]
- Commission Regulation (EC) no. 1998/2006 of 15 December 2006 on the application of Articles 87 and 88 of the Treaty to *de minimis* aid.[31]
- Commission Regulation (EC) no. 800/2008 of 6 August 2008 declaring certain categories of aid compatible with the common market in application of Articles 87 and 88 of the Treaty (block exemption regulation).[32]

---

[23] Cit. Roberti (1997), p. 17.

[24] Published in the Official Gazette no. L 142 of 14 May 1998. Regulation no. 994/98 authorizes the Commission to declare (pursuant to art. 87 of the Treaty) that some categories of aid are compatible with the common market and not subject to the notification requirement (provided for by Article 93, paragraph 3 of the Treaty). For our purposes here, it is of interest that among the types of aid that the Commission may declare compatible under particular conditions with the common market is also aid for small and medium-sized enterprises.

[25] Published in the Official Gazette of the European Union no. L 83 on 27 March 1999.

[26] Published in the Official Gazette of the European Union no. L 10/20 of 13 January 2001.

[27] Published in the Official Gazette of the European Union no. L 10/33 of 13 January 2001.

[28] Published in the Official Gazette of the European Union no. L 140 of 30 April 2004.

[29] Published in the Official Gazette of the European Union no. L 308/5 of 24 November 2009.

[30] Published in the Official Gazette of the European Union no. L 302/29 of 1 November 2006.

[31] Published in the Official Gazette of the European Union no. L 379/5 of 28 December 2006.

2.5. The Discipline Applicable to Aid Granted to an SME

As said, of close relevance to our discussion here are the regulations that apply to aid granted by a Member State to a small and medium-sized enterprise. For the definition of 'small and medium-sized enterprise' (SME),[33] we refer to the Commission Recommendation of 6 May 2003[34] concerning the definition of micro, small and medium-sized enterprises. This Recommendation updates – in order to give better definition to the economic reality of SMEs – the definition contained in the Commission Recommendation (96/280/EC) of 3 April 1996 also concerning the definition of small and medium-sized enterprises.[35] It is clear that the main objectives of the two Recommendations are to have common European-level rules and definitions, and to ensure that certain enterprises, whose economic power outweighs that of a SME, are not able to take advantage of support mechanisms specifically intended for SMEs.

What criteria are considered by the Commission? It is by now well known that enterprises are defined by their staff and their turnover, i.e. their total balance sheet. The 2003 Recommendation therefore defines a 'medium-sized enterprise' as one which employs fewer than 250 persons and whose turnover does not exceed 50 mln euros or whose annual balance sheet does not exceed 43 mln euros. Instead defined a 'small enterprise' is one employing fewer than 50 persons and whose turnover or total annual balance sheet does not exceed 10 mln euros. Finally, defined a 'micro enterprise' is one employing fewer than 10 persons and whose turnover or total annual balance sheet does not exceed 2 mln euros.[36]

That said, it should be pointed out that SME support measures[37] which involve state aid as per Article 107, I c., of the Treaty must comply with the procedures relative to state aid described earlier. There are some exceptions, however.

We consider first the '*de minimis*' aid currently disciplined by Regulation no. 1998/2006.[38] What does this rule provide for? The already-cited Council Regulation (EC) no. 994/98 empowers the Commission to set out in a regulation a threshold below which aid is deemed not subject to the notification procedure because it does not meet all the criteria stated by Article 107, I c. Thus the Commission, first with Regulation (EC) no. 69/2001 and then with Regulation no. 1998/2006, establishes the total amount of '*de minimis*' aid allowed.

What is the aid threshold below which Article 107 of the Treaty can be considered inapplicable? The '*de minimis*' rule states that aid measures of up to 200,000 euros per firm,[39]

---

[32] Published in the Official Gazette of the European Union no. L 214/3 of 9 August 2008.

[33] On this see European Commission (2009) p. 3, where one reads that small and medium-sized enterprises (SMEs) are the 'backbone' of the European economy. There are more than 23 mln SMEs in the EU representing 99% of businesses.

[34] Published in the Official Gazette of the European Union no. L 124 of 20 May 2003.

[35] Published in the Official Gazette of the European Union no. L 107 of 30 April 1996.

[36] To be noted is that Article 3 of the 2003 Recommendations identifies three types of enterprises for the purposes of calculating staff numbers and financial amounts: autonomous, partner, and linked, according to their relations with other enterprises in terms of capital participation, voting rights, and right to exercise dominant influence.

[37] On this see Pugliese (1976), p. 1124; Biglieri (1992), p. 1387; Capelli and Gratani (1995), p. 589; Franchini (1995), p. 2283; Palmieri (2002), p. 416; Pedilarco (2005), p. 383; Benvenuto (2005), p. 2213; Ferraro (2010), p. 335.

[38] See Council Regulation (EC) no. 1998/2006 of 15 December 2006 on the application of Articles 87 and 88 of the Treaty to *de minimis* aid, published in the Official Gazette of the European Union no. L 379/5 of 28 December 2006.

[39] The amount of 100,000 euros (over a three-year period) specified by Regulation (EC) no. 69/2001 was increased to 200,000 euros by Regulation no. 1998/2006.

over three fiscal years, do not constitute state aid within the meaning of the Treaty. Hence Member States may grant aid of that amount without procedural obligations.

The '*de minimis*' rule is therefore an instrument that allows Member States rapidly to grant subsidies of small amounts to enterprises (and particularly to SMEs) because they are not subject to the notification requirement. Under Article 2, I c., of the Regulation rule in question, in fact, aid that fulfils the conditions stated in the same Article must be considered aid that does not meet all the criteria of Article 107 of the Treaty and is therefore not subject to the notification requirement.

Let us now consider incentives that constitute state aid but can be granted directly by Member States without notification to the Commission. The reference is to the regime established by Commission Regulation (EC) no. 800/2008 of 6 August 2008.[40] This rule, the so-called "block exemption regulation", notably simplifies the rules on state aid by specifying categories of aid[41] which if they fulfil the conditions specified in the Regulation are exempt from the notification requirement. Consequently, Member States can grant such categories of aid immediately, and only subsequently inform the Commission.

We finally consider the state aid deemed compatible if it fulfills the conditions of the Commission's relative intentions and rules. The reference, of course, is to aid measures to support the growth and the development of SMEs;[42] aid for research, development and innovation, aid for environmental protection, regional aid, aid for SME investments and employment, training aid, etc.

2.6. New Developments that May Be Identified in the Discipline of State Aid

In conclusion to this brief survey we should mention some important and recent novelties in the regulation of state aid.

Firstly, we investigate the effect of the programme to re-launch[43] the Lisbon strategy, which in fact attributes an important function to the control of state aid and 'suggests' a new role for it.[44]

Secondly, mention should be made of the impact that the current financial and economic crisis[45] has had on the regulation of state aid.

---

[40] We refer to Commission Regulation (EC) no. 800/2008 of 6 August 2008 declaring certain categories of aid compatible with the common market in application of Articles 87 and 88 of the Treaty (General block exemption Regulation), published in the Official Gazette of the European Union no. L 214 of 9 August 2008.

[41] Under Article 1, I c.., the regulation applies to the following categories of aid: (a) regional aid, (b) aid for SME investments and employment, (c) aid for firm start-ups by female entrepreneurs, (d) aid for environmental protection, (e) aid to SMEs for consulting services and participation in trade fairs, (f) aid in the form of risk capital, (g) aid for research, development and innovation, (h) aid for training, (i) aid for disadvantaged and disabled workers.

[42] On this see European Commission (2009).

[43] The European Council meeting of March 2000 held in Lisbon set the strategic goal for Europe of becoming "the most competitive and dynamic knowledge-based economy in the world capable of sustainable economic growth with more and better jobs and greater social cohesion". In 2005, when it was apparent that these objectives had not been achieved, the EU Member States decided to re-launch the Lisbon strategy, concentrating efforts on the two main objectives of economic growth and employment.

[44] On this see Tufano (2011), p. 35, where we read "incentives to enterprises have assumed a new role, being understood as key instruments for the solution of market failures (regional imbalances, firms and sectors in crisis) and incentives for desirable behaviours (such as investments in environmental protection, research and development, etc., therefore as efficacious levers of development)".

[45] On this see Rabitti Bedogni (2010) and Tosato (2011), p. 11.

As regards the Lisbon strategy, it has been correctly pointed out that "the Commission has significantly modernised[46] the state aid rules in order to encourage Member States to target public support better on sustainable investments, thus contributing to the Lisbon Strategy for growth, jobs and competitiveness".[47]

The European Commission has taken an important step towards more general reform of the rules and mechanisms regulating state aid to national enterprises with adoption of the 2005 Action Plan for the state aid sector.[48] In this Action Plan,[49] the Commission sets out a roadmap for reform of state aid for the years 2005-2009, not only illustrating the main objectives of the reform but also indicating the instruments with which to achieve those objectives. The proposed radical change of the rules (both substantial and procedural) on state aid is based on the following four elements:[50]

i) less and better targeted aid;
ii) a more refined economic approach;
iii) more effective procedures, better enforcement, greater predictability, and enhanced transparency;
iv) shared responsibilities between the Commission and Member States.[51]

The ten-year cycle of the Lisbon strategy concluded in 2010. The European institutions and Member States then began discussion in order to verify the results achieved, to identify the obstacles encountered, and to define a 'post-2010' strategy. The new 'Europe 2020' strategy adopted on the occasion of the European Council of the Heads of State and Government of 17 June 2010 set out three priorities to pursue: smart growth (developing an economy based on knowledge and innovation); sustainable growth (promoting a more resource-efficient, greener, and more competitive economy); and inclusive growth (promoting a high-employment economy favouring social and territorial cohesion).[52] The Commission also set the following EU targets: 75% of people aged between 20 and 64 in employment; 3% of the EU's GDP invested in R&D; the '20/20/20' climate/energy targets fulfilled (including a 30% increase of emissions reduction if conditions allow it); the school drop-out rate should be less than 10%, and at least 40% of young people should be graduates; 20 mln less people should be at risk of poverty.[53]

In regard to our discussion here, to be noted is that the new Strategy devotes considerable space to 'state aid'. The Commission Communication relative to "Europe 2020" of 3 March

---

[46] To be stressed in this regard is that the European institutions make increasing reference to 'modernization'. The expression 'modernization of the economy' is frequently used to denote the neoliberal doctrine of recent decades advocating the state's withdrawal from the economy and the opening of markets to competition.

[47] Cit. European Commission (2009), p. 5.

[48] Commission of the European Communities, state aid action plan, Less and better targeted state aid : a roadmap for state aid reform 2005–2009, Brussels, 7 June 2005, COM (2005) 107 (Consultation document) final.

[49] On this see Schepisi (2011), p. 17.

[50] Point 18 of the Action Plan. For a description of the four elements see http://europa.eu/legislation_summaries/competition/state_aid/l26115_it.htm.

[51] The Commission stresses that it cannot improve the rules on state aid without the effective support of Member States and their full commitment to complying with the requirement to notify all aid that they intend to grant and to apply the rules correctly.

[52] Cit. Communication from the Commission, Europe 2020, A strategy for smart, sustainable and inclusive growth, Brussels, 3.3.2010, COM(2010) 2020.

[53] Cit. Communication from the Commission, Europe 2020, A strategy for smart, sustainable and inclusive growth, Brussels, 3.3.2010, COM(2010) 2020.

2010, in fact, stresses that: "State aid policy can also actively and positively contribute to the Europe 2020 objectives by prompting and supporting initiatives for more innovative, efficient and greener technologies, while facilitating access to public support for investment, risk capital and funding for research and development".[54]

Turning to the current situation of crisis, to be emphasised is that the Commission has expressed willingness to review its position in consideration of the severe economic and financial crisis[55] affecting Countries. The Commission foresees that the current crisis will impact – albeit in different ways and to various extents – on all Member States, very probably causing an increase in unemployment, a fall in demand, and a deterioration of public finances. Consequently, given the severity of the economic and financial crisis, the Commission – with its Communication on a "Temporary Community framework for state aid measures to support access to finance in the current financial and economic crisis" (2009/C 83/01)[56] – has decided that certain categories of state aid[57] may be justified (temporarily and in order to remedy SME difficulties) and may be declared compatible with the common market on the basis of the already-cited derogation provided for by art. 107, III c., clause b of the TFEU: "aid to promote the execution of an important project of common European interest or to remedy a serious disturbance in the economy of a Member State".

2.7. 'State Aid' in the Recent Experience of Certain European Countries

A concise description of the quantitative scenario concerning the effective granting of state aid in the largest Countries of the European Union is made possible by its constant monitoring – since observation began on the granting of aid as a practice forbidden, or at any rate discouraged, in the EU – by the Directorate General of the European Commission which deals with issues concerning competition, (Brancati and Maresca, 2010, p. 260). During the 1990s and 2000s, the absolute and relative (with respect to GDP) situations of Countries changed greatly following trends recorded especially in Spain and Italy: resources allocated to non-agricultural productive activities were substantially reduced, and relative positions also changed with respect to the other reference Countries. "The degree of financial intervention for enterprises in Italy was among the lowest in Europe even before the measures intended to reduce the effects of the crisis and presumably less than, for instance, in the USA,[58] where there is massive intervention not only through the use of public demand (military spending, research. etc.) but also with direct outlays which are difficult to calculate because they often made by individual states or counties"[59] (MET, 2010, pp. 260-1).

Moreover, in the last decade the demand for public grants increased in other Countries, assuming different features: for example, the extraordinary effort required of enterprises,

[54] Cit. Communication from the Commission, Europe 2020, A strategy for smart, sustainable and inclusive growth, Brussels, 3.3.2010, COM(2010) 2020.

[55] On this see Ferraro (2010), p. 335 and also De Andreis, Maré and Perissich (2009).

[56] Published in the Official Gazette of the European Union no. C 83/1 of 7 April 2009.

[57] The categories foreseen by the Communication from the Commission, no. 2009/C 83/01, are the following: aid in limited and compatible amounts; aid granted in the form of guarantees; aid in the form of subsidised interest rates; aid for the production of 'green' products; and risk capital measures.

[58] Unfortunately, systematic and complete data are lacking for the USA.

[59] As in Italy, where also regional administrations, provincial administrations, and Chambers of Commerce allocate public incentives to enterprises situated in their geographical areas of competence.

particularly SMEs, to adapt to globalization has induced some governments to increase measures. Overall, therefore, the flow of resources allocated to state aid in the EU has tended to increase, though only to a slight extent. Instead, if only the five principal Countries are considered, appreciable decreases are apparent (obviously excluding anti-crisis measures): the modest fall recorded by Germany is accompanied by a further decrease in Spain and Italy. By contrast, France records substantial growth.

With regard to the average value of state aid relatively to the EU-27 aggregate, the measures adopted in the period 2007-08 (again net of anti-crisis measures) manifest high intensity in Spain and Germany, placing them above the European average.[60]

By contrast, France is positioned below the average value, together with Italy, whose flow of aid is particularly modest.

If the various types of subsidies are considered,[61] one notes that in 2008 the capital grant, in its direct form or in that of tax deduction, was the instrument with the highest incidence, representing more than 90% of overall expenditure at EU-27 level. Italy, however, clearly records a greater relative weight of capital grants in comparison with the other four Countries.

The other types of subsidies, therefore, have a relatively minor weight, although Italy and Spain record a certain frequency of subsidised loans or interest facilitations. Guarantees and equity participations, however, are negligible in all Countries.

Table 1. State aid to industry and services (net of the financial sector) in certain Countries of the EU. % of GDP. Years 2009 and 2010

Country	2007-08 average	2009	2010	Var. 2009-10
UK.	0.22	0.22	0.24	+ 0.02
Italy	0.27	0.33	0.22	- 0.11
Spain	0.77	0.57	0.41	- 0.04
Germany	0.57	0.58	0.61	+ 0.03
France	0.38	0.69	0.66	- 0.03
EU 27	0.41	0.51	0.49	- 0.02

Source: our calculations on MET and European Commission - D.G. Competition data.

Table 2. State aid distinguished by financial instrument (net of anti-crisis measures). % Composition. Year 2008

Country	Capital grants	Tax exemptions	Equity Participations	Interest subsidies	Tax deferral	Guaran tees
France	44.8	52.1	0.0	2.9	0.0	0.2
Germany	42.6	51.4	0.3	3.5	-	2.2
Italy	76.5	14.8	0.1	8.6	-	0.0
UK	45.0	53.9	1.1	0.0	-	0.0
Spain	57.5	35.3	0.0	7.1	0.0	0.0
U.E. 27	51.7	42.8	0.5	3.3	0.1	1.6

Source: our calculations on MET and European Commission - D.G. Competition data.

[60] See Table 1.

[61] See Table 2, in which the values are expressed, following suitable financial conversions. in gross grant equivalents in order to show the aid content of the various measures and make the various types of subsidy comparable.

When the situation is updated to the final two years of the past decade,[62] once again there is evidence of diversified trends among the principal EU Countries.

Apart from the United Kingdom, where the share of state aid in GDP remains substantially unchanged, Italy records a tendency to contraction, unlike France and Germany, where the contrary is instead the case.

This finding – bearing in mind that the data are purged of bank bail-out measures – is of particular significance because it highlights that enterprises in the two economically strongest European Countries obtain resources practically double those of two Countries – Italy and Spain – which are in more difficult situations, also in terms of output and employment.

To conclude the analysis, there follow some brief critical comments concerning Italy in particular. In general, the predominant position in the Italian and European literature on aid for enterprises is based on arguments inspired by highly simplified good principles, according to which distortionary and harmful measures like those for industry are pointless; instead, it is essential to have a lean and efficient state able to guarantee general access to public goods, and delivering high-quality services.[63] In parallel, with respect to the public positions just outlined, enterprise policy measures continue to be adopted with not large, nor negligible amounts – and without particular attention paid to the way in which they are implemented or, even less, to the possibilities of their improvement. If this process continues, the entire ‘toolkit’ of industrial policies will be exhausted, and there will be a lack of convincing and consistent ideas about what strategy to implement, with instead more frequently emergencies or sporadic and unstructured solutions. The contradictions of policy tend to be all the more apparent when one considers that a predominant part of enterprise aid policies are now targeted on incentives for private research and the introduction of technological innovations. These are generally accepted objectives which are mostly pursued by instruments that are usually suggested by industrial policy (Brancati and Maresca, 2010, p. 267).

3. EU Policy for Enterprises: An Overview

3.1. EU Policy for Enterprises, with Special Reference to SMEs: General Aspects

The purpose of the EU’s policy for enterprises is to maintain, and if possible increase, the competitiveness of the latter with respect to their non-European rivals, as well as to create new employment, paying special attention to the needs of manufacturing industry, on the one hand, and small medium-sized enterprises (SMEs) on the other.[64]

European enterprise policy aims mainly to create an environment conducive to investments, not only for strategically important sectors like the aerospace industry and biotechnologies but also for more traditional ones, including the textiles industry and car

[62] See again Table 1.

[63] See Banca d’Italia (2009), which contains numerous studies reaching conclusions similar to those in the main text.

[64] In fact, despite the rapid growth in recent decades of the services sector, manufacturing industry is still the driver of the European economy, given that it accounts for 75% of total exports by the EU. For a general analysis of the role performed by SMEs at European level see European Commission (2010), while for an analysis of the importance assumed by SMEs in the OECD member-Countries, even in times of crisis like 2008-09, see OECD (2010).

manufacturing, where it is crucially important for companies to adopt advanced technologies. Hence, in the near future, the growth of the EU will very probably be driven by innovative, knowledge-based sectors, which nevertheless require a healthy industrial system and the means with which to exploit the potential of emerging technologies. This entails closing the current gap between universities and enterprises so that winning ideas developed in the laboratory can be converted into successful products.

Since there are 23 mln SMEs, representing 99% of all undertakings in Europe, they constitute a key engine for economic growth, innovation, employment, and social integration. The European Commission aims to promote successful entrepreneurship and to improve the conditions for SME activities, enabling them fully to realize their potential in the global economy. The European Commission concerns itself with various policies for entrepreneurs and SMEs in Europe, as well as mechanisms to support them (networks, business support measures). Moreover, it works for both existing and potential entrepreneurs, devoting particular attention to specific forms of entrepreneurship, such as female-run businesses, crafts firms, or social economy enterprises. All actions to support SMEs and entrepreneurship have been brought within a single framework, the *Small Business Act* for Europe (SBA), which Member States have committed themselves to implementing together with the European Commission. Since SMEs should be supported at local level, the Commission's activities also consist in helping Member States and the regions to develop policies to promote entrepreneurship, to improve the situation of SMEs throughout their life-cycles, and to assist them in gaining access to new markets. Also the identification and exchange of good practices are essential elements of such policies (Lopriore 2009).

Although the world of industry is often associated with large multinational companies, in fact the majority of European undertakings (92%) are small firms with fewer than 10 employees, especially in some Countries of southern Europe like Italy, or in the less developed European regions.[65] Since micro enterprises account for two-thirds of jobs in the EU, it is obvious that they receive particular attention in the programmes devised for funding enterprises at European level, among which:

- the Competitiveness and Innovation Programme, for which an investment of 3.6 bln euros is foreseen in the period 2007-13, especially in energy efficiency, renewable sources, and information/communication technologies;[66]
- the Seventh Framework Programme (7th FP), the EU's main programme for funding scientific research (in universities and industry), with a budget amounting to more than 7 bln euros per year.[67]

[65] For a specific analysis of the instruments, like tax and aid policies, devised to promote innovative regional systems see Bruno (2009), which shows – amongst other things – that in recent years such instruments have undergone a marked process of revision and restructuring, also in regard to the principles laid down by the EU regulations on the structural funds used within the framework of the cohesion policy. More specifically, the European Commission has induced national authorities to abandon the most traditional and consolidated aid regimes in favour of both forms of support for particular sectors (environmental protection, research, development and innovation, etc.), and modern financial engineering solutions, i.e. those that presuppose close complementarity among the various forms of incentives, with the purpose of remedying the shortcomings complained about by SMEs.

[66] For more detailed analysis of this programme see section 3.3.

[67] See section. 3.3. To be mentioned here, however, is that, also in 2011, especial attention was paid to SMEs, since they were allocated, amongst other things, a package amounting to almost one bln euros. Because SMEs are considered crucial for innovation, they will receive particular incentives to participate in the programme. Not

Besides promoting entrepreneurship and skills, the aim of the EU is to improve access by SMES to markets and to enhance their research and development capabilities. Moreover, SMEs in need of specific information and consultancy can obtain them via the Enterprise Europe Network, which has around 500 one-stop shops throughout the EU and is partially financed by the latter. But also the reduction of bureaucratic paperwork is one of the EU's maximum priorities[68]. Finally to be mentioned is that the Commission issues an Enterprise Finance Index which facilitates the search for facts and figures on the development of SME finance in the various Countries of the EU.

3.2. The EU's Policy for SMEs

In order to support SMEs – which, as said, perform a decisive role in the competitiveness and dynamism of the European economy – in realization of their growth potential, the EU has for some time undertaken various actions,[69] in particular, the promotion of entrepreneurship and the creation of a commercial environment better suited to the needs of small enterprises. More specifically, the lines of action for SMEs currently pursued by the European Commission are the following:

- promoting entrepreneurship;
- improving the business environment in the various phases of a firm's life cycle (start-up, management, and growth);
- facilitating access to markets;
- fostering the diffusion of best practices;
- and, finally, creating a more systematic set of measures denominated the "Small Business Act (SBA)" for Europe (European Commission - DG Enterprise and Industry 2008).

In what follows, the last of these lines of action will be treated in particular, because it indubitably represents the most systematic form of intervention recently adopted at EU level.[70]

Adopted in June 2008, the "Small Business Act for Europe" (SBA) reflects the Commission's political will to recognise the central role of SMEs in the EU economy and for

only will they benefit from simplified rules, but the package of investments of almost one bln euros includes pilot-actions for SMEs both in regard to health and a specific financing mechanism whereby risks are shared with the European Investment Bank (EIB). For a statistical analysis of the data on the contributions allocated to SMEs within the 7th FP until 1 April 2011see European Commission. DG Research and Innovation (2011), from which study it emerges that, as of that date, the objective of granting to SMEs 15% of the total resources allocated among the various sub-programmes had already been almost achieved (14.4%).

[68] This last point should obviously be developed further, but here it is simply mentioned while the focus is on incentives of a financial nature.

[69] For the policy pursued until the end of the 1990s see Calabrese (2001), while for the main measures adopted in the past decade, among them the European charter for small enterprises and the new definition given to SMEs, see Sediari and Dell'Aquila (2007).

[70] The elements we are about to present have been drawn mainly from the European Commission's portal, where all information supplied by the EU about and for the benefit of SMEs is collected: see http://ec.europa.eu/small-business/index_en.htm. As a result, the topics dealt with here are described as briefly and schematically as possible, mainly because the underlying philosophy has already been illustrated.

the first time it puts into place a comprehensive SME policy framework for the EU and its Member States.

It aims to improve the overall approach to entrepreneurship, permanently anchor the 'Think Small First' principle in policy making from regulation to public service, and to promote SMEs' growth by helping them tackle the remaining problems which hamper their development.

The Small Business Act for Europe applies to all independent companies which have fewer than 250 employees: 99% of all European businesses.

In short, the SBA:

- is a set of 10 principles which should guide the design and implementation of policies both at EU and national level. This is essential to level the playing field for SMEs throughout the EU and improve the administrative and legal environment so as to allow these enterprises to unleash their full potential to create jobs and growth;
- is an ambitious package of concrete and far reaching new measures including four legislative proposals which translate these principles into action both at EU and Member State level;
- was endorsed politically by the EU Council of Ministers in December 2008 to ensure the full commitment of both the Commission and the Member States together with regular monitoring of its implementation.

On November 2011, the Commission presented a new approach to ensure that the EU responds better to the needs of small businesses. From now on, the European Commission will seek wherever possible to exempt micro-enterprises from EU legislation or introduce special regimes so as to minimise the regulatory burden on them.

In a report to the Council and the European Parliament, the Commission presented a list of initiatives of this kind already taken and to be examined for the future. It announced stronger means to ensure the input of micro-enterprises and small and medium-sized enterprises (SMEs) to the formulation of new EU initiatives.

The SBA Review presents an overview of the progress achieved in implementing the Act, and sets out new actions to respond to challenges resulting from the economic crisis. Between 2008 and 2010, the Commission and EU Member States took measures to ease the administrative burden on small businesses, to facilitate SMEs' access to funding, and to support their access to global markets. Although most initiatives foreseen by the SBA have been initiated, the review reveals that more must be done to help SMEs.

3.2.1. What Does the SBA Do for SMEs?

The Small Business Act aims to:

I) Design Rules according to the 'Think Small First Principle'

All new legislative and administrative proposals at European and national level should be subjected to an 'SME test' to assess their impact on small businesses. Where this impact is considered negative, Member States will be able to use measures such as derogations, transition periods and exemptions. New legislation can include fewer reporting requirements for SMEs and measures to cut down on paperwork. For example, through the EU Directive on

VAT invoices, electronic invoices are put on an equal footing to paper invoices. A separate EU Directive will offer the opportunity to EU Member States to relieve micro enterprises from the obligation to present their annual accounts.

II) Facilitate SMEs' Access to Funding

The European Investment Bank Group should increase the range of financial products that it offers to SMEs, particularly mezzanine finance. In addition, more funds will be made available by the Commission for micro-credit, and access to cross-border venture capital will be facilitated.

Moreover, the new Directive seeks to combat late payments in commercial transactions, as these can be crippling for SMEs.

III) Help SMEs to Get Most out of the EU's Single Market

A new statute for a European Private Company will allow SMEs to start up and operate according to the same company law provisions throughout Europe.

The new legislation on lower VAT rates for labour intensive services will have a positive effect on SMEs in sectors such as hairdressing, catering, and repair services.

IV) Create an Entrepreneurial Environment

With 'Erasmus for young entrepreneurs', new and budding entrepreneurs can gain experience and insight by spending time in an SME in a different Country.

Honest entrepreneurs who have faced bankruptcy should quickly be given a second chance, by promoting a positive attitude in society towards a fresh start.

As part of the European SME Week 2009 and 2010 more than 2,000 events were organised throughout the EU to highlight business support measures and new opportunities. These special weeks have also helped to raise awareness about the important contribution that entrepreneurs make to the EU economy.

V) Adapt Public Policy Tools to SME Needs

A new code of best practice for public procurement will facilitate SMEs' access to public procurement contracts. The code improves access to information about public contracts and how to bid on-line, reduces excessive financial requirements, and cuts down on paperwork.

Member States will have scope for providing direct assistance to SMEs. A higher level of state aid to smaller companies will be allowed, and rules on state aid to SMEs will be made simpler. In addition, national governments will no longer need to notify this support to the Commission.

3.2.2. The Think Small First Principle

The Think Small First principle requires legislation to take SMEs' interests into account at the very early stages of policy making in order to make it more SME-friendly. A range of tools are available to ensure the effective implementation of the principle. These include the application of an SME test to forthcoming legislative proposals, the use of specific SME provisions in legislation in order to avoid a disproportionate burden on SMEs, the consultation of the SME stakeholders, the work of the SME Envoy, the use of Common Commencement dates for legislation relevant for business, etc.

The Think Small First principle is at the heart of the Commission Communication "Small Business Act for Europe". The overarching objective of this political document is to irreversibly anchor the Think Small First principle in policy-making through the application of 10 politically binding principles and a series of actions to be performed by the Commission and the Member States.

Report of the Expert Group Think Small First – Considering SME Interests in Policy-Making

As part of the efforts to implement the Small Business Act for Europe and in accordance with its mission to promote the sharing of knowledge and experience among the Member States, in 2008 the Commission formed an expert group from national administrations to consider the application of the Think Small First principle at EU and national level. The group's report issued in March 2009 provides a definition of the Think Small First principle and presents a number of good practices at national and EU level on how to Think Small First in order to implement the Small Business Act for Europe. The purpose of highlighting European good practice is to suggest to policy makers in the Member States concrete solutions for the improvement of the business environment.

3.2.3. Consultation of SME Stakeholders

Consulting SME stakeholders is indispensable for the implementation of the Think Small First principle. The Report on the consultation of stakeholders in the shaping of small business policy at national/regional level provides examples of European best practice. The Commission uses a variety of tools in order to reach European SMEs. These include regular consultations with sectorial and horizontal business organisations represented at EU level, consultations of panels of businesses, and online public consultations.

SME Envoy

The European Commission's SME Envoy is the contact point for small businesses and SME stakeholders at the European Commission, giving them the chance to provide feedback on regulation, but also to be involved at an early stage in the conception of new policies and laws. The Envoy is a high-ranking official and serves as Deputy Director General at DG Enterprise and Industry. Within the Commission the Envoy plays the role of ambassador for SMEs' interests and raises awareness about SMEs' problems, as well as promoting less burdensome regulation for the businesses.

Common Commencement Dates

The concept of "Common Commencement Dates" is an innovative solution to improve the lives of SMEs. In line with the Think Small First principle and the Better Regulation Agenda, the idea is that business-related legislation comes into force on a limited number of fixed dates, with a view to ensuring a more streamlined flow and early information on new regulation concerning SMEs.

Listening to SMEs

The opinions of SMEs are important when developing new EU legislation or programmes that have a direct impact on their operations. This is why the Commission, through the

Enterprise Europe Network, operates two mechanisms to test the impact of EU legislation and programmes.

In Advance via SME Panels

SME Panels are organised to consult SMEs about forthcoming EU legislation and policies. The Enterprise Europe Network partners select suitable SME participants, run the SME panels and provide the Commission with the results. The result are used when preparing new legislative or policy proposals. SME panels are operated in liaison with the SME Envoy to ensure that the needs of SMEs are continuously taken into account in EU law-making.

Retrospectively via the SME Feedback Mechanism

The SME feedback mechanism allows the Enterprise Europe Network partners to collect the views and feedback from SMEs on a broad range of EU policy initiatives, actions, legislation or programmes related to the internal market.

Typical policy areas where feedback will be requested include the environment, sustainability, employment and social affairs, innovation support, taxation and customs and, in more general terms, better regulation and simplification.

3.2.4. Decision Making Process

The policy-making process in the European Union sees legislative proposals from the Commission debated both by Members of the European Parliament and by representatives of Member States' governments within the European Council. And in general, the Commission consults with interested parties both formally and informally before it draws up its proposals, such as the Small Business Act for Europe. Hence there are opportunities for those affected (or potentially affected) by policy to make their views known at several stages in the process.

For SMEs – often the largest group of enterprises to be affected by new policy – there are clear difficulties in putting their views across to policy-makers. SMEs do not usually have the resources which large firms can dedicate to influencing policy-making or lobbying. For this reason, the Commission has appointed an SME Envoy and encourages an effective and wide-ranging consultation of SMEs as one element of its Think Small First principle.

The Commission also encourages Member States and regions to adopt similar measures, ensuring greater representation of SME interests at all levels of policy-making. To this end, it has sought to identify and share examples of effective SME consultation and involvement in policy-making.

3.2.5. Implementation of the "Small Business Act" for Europe

A first report on the "Small Business Act" for Europe was adopted by the European Parliament on March 2009. On December 2009, the Commission adopted a report which highlighted the progress made in implementing the SBA, both at EU and national level. On February 2011, the Commission issued a Communication taking stock of progress achieved and proposing new actions to be implemented at EU and national level.

The Member States

The Member States have committed themselves to implementing the SBA by adopting the "Competitiveness Council Conclusions of 1 December 2008".

The SBA Action Plan, annexed to the Competitiveness Council Conclusions, focuses on short and medium-term measures to improve SMEs' access to finance, to improve the regulatory environment and to enhance market access for SMEs. These priority measures are an integral part of the Commission's response to alleviate the negative impact of economic and financial crisis on enterprises, especially SMEs.

The European Council on 11-12 December 2008 supported the full implementation of the Action Plan for the SBA adopted by the Competitiveness Council on 1 December 2008.

The European Economic Recovery Plan, adopted by the Commission on 26 November 2008, also includes far-reaching measures to enhance access to finance for SMEs, to reduce their administrative burden, and to promote entrepreneurship.

Since 2009, the Member States have been expected to report on their implementation measures in their Lisbon National Reform programmes.

The Competitiveness Council of 28 May 2009 recognised good progress in implementing the SBA. All legislative proposals are now on the table and the SME test is applied in the Commission impact assessment procedure. Member States have agreed unanimously on the importance of strengthening action to improve access to finance and decrease the regulatory burden on businesses.

3.2.6. SME Performance Review

The SME Performance Review represents one of the main tools employed by the European Commission to monitor and assess Member States' performances in implementing the Small Business Act. Launched by the European Commission in 2008, it represents a comprehensive source of information on the performance of SMEs in Europe and uses a wide range of success indicators, focusing most notably on the measures in the SBA Action Plan.

SMEs' contribution to the EU's prime economic objectives is acknowledged and well documented in both the Lisbon strategy for economic growth and more and better jobs and its successor Europe 2020 strategy. One step further, the Small Business Act for Europe (SBA) establishes a comprehensive SME policy framework for the EU and its Member States. In the review of the Small Business Act (SBA) in February 2011 the Commission and the Member States acknowledged that strong governance is key for successful implementation of the SBA.

The SME Performance Review comprises an Annual Report complemented by SBA fact sheets and a set of studies providing in-depth information on issues that are particularly important to small and medium-sized enterprises in Europe.

3.2.7. SME Test

The idea behind an SME Test is to analyse the effects of a legislative proposal on SMEs. The SME Test is part of the regulatory impact assessment.

Regulatory Impact Assessments

Impact assessment (IA) analyses the likely economic, environmental and social effects of legislation by assessing the costs and benefits of policy options. This highlights possible unintended consequences, synergies and trade-offs, and helps to reach balanced, sustainable decisions.

The European Commission Impact Assessment Guidelines set out the methodology used for the Commission Impact Assessments. The SME Test is presented in the annexes to the Guidelines.

Applying the Think Small First Principle in EU Legislation and Programmes

The SME Test – as developed in the Commission impact assessment guidelines – was inspired by a 2008 study "Evaluation of the application of the Think Small First principle in EU legislation and programmes". The findings of the study highlighted the lack of a coherent approach to the evaluation of the impacts on SMEs and proposed a Think Small First toolkit which fed into the Commission SME Test methodology. The toolkit in particular provides guidance on how and in what cases it is appropriate to introduce specific measures for SMEs, such as exemptions, reduced fees, and simplified reporting, to avoid a disproportionate burden on SMEs.

The Commission 'SME Test'

The SME Test comprises three main steps:

i) Preliminary assessment of businesses likely to be affected;
ii) Measurement of the impact on SMEs (cost/benefit analysis);
iii) Use of mitigating measures, if appropriate.

The opinion of business representatives feeds into the assessment of how a new policy could affect industry. In addition to ongoing consultation with stakeholders, the Commission has developed a number of tools that help to obtain the opinion of businesses.

SME organisations are given between eight and twelve weeks to provide written feedback to the Commission. This gives time for representative groups to consult their members on often technical legislative proposals, analyse their responses and, in some cases, build alliances with other trade bodies. Consultations are generally avoided during major holiday periods and, where this is not possible, extended deadlines can be given.

The Mitigating Measures

If, following the cost/benefit analysis, it is felt that SMEs would be disproportionately affected by a regulation, mitigating measures can be used to soften the impact on smaller companies. On occasion, it may not be possible to prioritise the interests of SMEs without compromising the main objective of a given proposal (which could be to protect human health or to safeguard the environment). In these cases it would be appropriate to think of introducing specific measures that might ease the burden on SMEs. These include:

- *complete or partial size-related exemptions for SMEs or micro-businesses.* (e.g. Commission proposal for exempting micro-companies from certain accounting requirements);
- *temporary reduction or exemptions* (e.g. some legislation in the field of health and safety or transport);
- *reduced fees* (e.g. REACH Regulation or European Medicines Agency);
- *simplified reporting obligations for SMEs* (e.g. statistics related to the trade of goods between Member States);
- *specific information campaigns or user guides, training and dedicated helpdesks /offices* (e.g. REACH national helpdesks).

Details of how such measures can be used, along with suggestions as to the circumstances where they are likely to be most effective and direct examples of their application in Community law, are found in the Think Small First Toolkit. This was developed as part of the framework study "Application of the 'Think Small First' principle in EU programme and legislation".

3.3. The main Financing Programmes for European SMEs

Access to finance is vital in order to start or expand a business, and the EU provides finance for small firms in different forms - grants, loans, and in some cases, guarantees. In addition, the EU funds specific projects. EU funding can be divided into two categories:

- direct funding through grants;
- indirect funding through national and local intermediaries (Ganslandt, 2009).

3.3.1. The Multiannual Programme for Enterprises and Entrepreneur-ship (MAP)[71]

Developed during the first half of the last decade was the *Multiannual Programme for Enterprises and Entrepreneurship* (MAP, which provided for three tools, of which one consisted of financial instruments). These financial instruments amounted to a total of 317 mln euros, which during the period between 2011 and 2005 were managed by the European Investment Fund (EIF) through national financial intermediaries.

By using these resources, three actions could be developed:

- a start-up mechanism;
- an SME guarantee mechanism; and
- start-up capital for innovative enterprises.

3.3.2. The Competitiveness and Innovation Framework Programme 2007-2013 (CIP)

Subsequently, the Competitiveness and Innovation Framework Programme 2007-2013 (CIP) was established with the aim of providing a response in line with re-launching the Lisbon strategy. It included within a common framework the specific community programmes and the pertinent parts of the other EU programmes in sectors fundamental for the purposes of promoting productivity, capacity, and innovation, as well as sustainable European growth.

It is complementary to the Seventh Framework Programme for Research and Technological Development[72] and contributes to closing the gap between research and innovation and to promoting all forms of innovation.

More specifically, the aim of the programme is to foster the competitiveness and innovative capacity of the Community, as an advanced knowledge society, with sustainable development based on robust economic growth and a strongly competitive social and market economy with a high level of environment protection and improvement.

---

[71] On these EU programmes for SMEs, see also MISE – Italia (undated) which also illustrates the main financing opportunities for SMEs arising out of EU external assistance programmes: that is, towards third Countries.

[72] See later in this section.

The Programme brings together activities which in the past were divided among the *Multiannual Programme for Enterprise and Entrepreneurship* (MAP), certain activities relating to industrial competitiveness, and elements of the LIFE+ Programme (in the environmental field).

The Framework Programme pursues the following objectives:

i) to foster the competitiveness of enterprises, in particular of SMEs;
ii) to promote all forms of innovation, including eco-innovation;
iii) to accelerate the development of a sustainable, competitive, innovative, and inclusive information society;
iv) to promote energy efficiency and new and renewable energy sources in all sectors, including transport.

The available financial resources amounted to 3,621 mln euros, which were issued directly by the DG Enterprise and Industry – European Commission.

The framework programme is open to participation – besides the Member States – by (potential) candidates for membership, which from 2007 – where permitted by agreements and procedures – were: Albania, Bosnia-Herzegovina, Croatia, the former Yugoslavian republics of Macedonia, Montenegro, and Serbia (including Kosovo), Turkey, Norway, Iceland, and Liechtenstein (from EFTA/EEA).

3.3.3. The Enterprise Europe Network

Within the Competitiveness and Innovation Framework Programme 2007-2013 (CIP), the DG Enterprise and Industry of the European Commission has instituted and funded the Enterprise Europe Network, which is an information and assistance network with the aim of assisting and advising SMEs on the EU's policies, programmes, and funding.

The network links around 500 organizations, including Chambers of Commerce, regional development agencies, and university research centers, as well as 4000 experts and professionals. The Enterprise Europe Network also extends beyond the borders of the EU to include Turkey, the former Yugoslavian republic of Macedonia, Norway, Iceland, Armenia, Israel, and Switzerland. It targets all research enterprises and public and private bodies, and offers business opportunities in Europe through the conclusion of technical collaboration agreements, production agreements with assistance and the transfer of know-how, joint venture agreements, agreements for the exploitation and/or assignment of patents, license agreements, purely commercial agreements, and partnership agreements for Community programmes such as the 6th Framework programme and the CIP.

The network partners provide information on access to European policies, programmes, and funding opportunities. They also:

- help companies to identify potential commercial partners, in particular in third Countries;
- help SMEs develop new products, gain access to new markets, and obtain information about EU activities and opportunities;
- provide consultancy to small enterprises on technical issues such as intellectual property rights and EU rules and legislation;

- act as channels of communications between business people and decision-making bodies at EU level;
- organise coopération among SMEs;
- promote innovation.

3.3.4. The Seventh Framework Programme for Research and Techno-logical Development (7FP)

This Programme is the European Union's main financial instrument for research. It was designed to re-launch European research, and to respond to the growing challenge of the USA and Asia within the framework of the renewed Lisbon strategy, which – as widely acknowledged – had the purpose of making Europe the most dynamic and competitive knowledge-based economy in the world by 2010 and able to achieve sustainable economic growth with more and better jobs and greater social cohesion.

It is different in some respects from the six previous Community research programmes, in that it has a larger budget, and concentrates on certain thematic areas; the budget for the 7th FP is 63% higher than that of the 6th FP at today's prices, which signifies greater resources for European research.

Overall, the available financial resources are 50,521 mln euros, with an annual budget of around 730 mln euros.

The main objectives of the 7th FP are:

i) to contribute to sustainable development through transnational cooperation in various fields corresponding to the main sectors of progress in knowledge and technology, so as to permit European research to acquire a leadership position in scientific areas and fundamental technologies;
ii) to promote a multidisciplinary approach and the creation of critical masses through the reinforcement of cooperation among universities, industries, research centres, and public bodies in the European Union as a whole and in the rest of the world;
iii) to foster the spread of knowledge and the transfer of results in all the thematic areas.

In addition to the Member States of the EU, the programme is open to the (potential) candidates, where permitted by agreements and procedures (Croatia, the former Yugoslavian republics of Macedonia, Serbia, and Montenegro, and Turkey), and to the following associated Countries: Iceland, Liechtenstein, Norway, Israel, and Switzerland.

Although the Programme mainly targets researchers from Member States, associated Countries, and European and international organizations, its fundamental value added consists in the high level of participation by enterprises, including SMEs.

It is based on four specific programmes:

a) the *Cooperation Programme*, which aims to stimulate cooperation and improve links between industry and research within a transnational framework. The objective is to build and consolidate European leadership in the most important research sectors;
b) the *Ideas Programme*, which is intended to incentivize exploratory research in Europe: that is, the discovery of new knowledge which fundamentally changes vision

of the world and way of life, the aim being to enhance European research excellence by promoting competition and risk-taking;

c) the *People Programme*, which mobilizes significant financial resources to improve the career prospects of researchers in Europe and to attract a larger number or high-quality young researchers. The Commission intends to encourage training and mobility in order to fully realize the human potential of European research;[73]
d) the *Capacities Programme*, which will give researchers effective tools with which to enhance the quality and competitiveness of European research. This means more investment in research infra-structure in the less efficient regions, in the creation of regional research clusters, and in research for the benefit of SMEs. This programme must also reflect the importance of international cooperation in research and the role of science in society.

3.4. The EU's Financial Institutions and Their Main Intervention Instruments in Favour of SMEs

The most important financial institutions of the European Union are:

a) the European Investment Bank (EIB);
b) the European Bank for Reconstruction and Development (EBRD); and
c) the European Investment Fund (EIF).

We will now briefly describe these institutions, with the emphasis on the role played by each in assisting SMEs, also through recently-activated tools such as Jeremie and Jasmine.

3.4.1. The European Investment Bank (EIB)

The EIB is the European Union's long-term credit institution. It grants loans to the public and private sectors for funding projects of European interest. It belongs to the Member States of the European Union, which jointly provide its capital according to shares which reflect the economic weight of each Member State within the EU. The EIB does not use funds from the EU budget; it funds itself by issuing loans on the financial markets.

The EIB is the bank of the European Union. Thanks to its Statute and shareholders – the 27 Member States – it has the best possible rating on the capital markets (AAA). The EIB can therefore borrow funds on the capital markets on favourable terms, which it passes on in the loans that it grants, in particular to SMEs through intermediary banks. These banks undertake to inform the SMEs (for example, by an individual information letter) of the EIB's involvement and its impact on the financial terms of the loan received.

The Bank invests in carefully-selected projects based on the following criteria:

i) they must be related to the achievement of the general objectives of the EU;
ii) they must be economically, financially, technically, and environmen-tally valid; and
iii) they must contribute to attracting other sources of funding.

[73] The programme in question exploits the success of the Marie Curie Actions, which have for many years offered mobility and training opportunities to European researchers.

The EIB also promotes sustainable development in Countries which are candidates, or potential candidates, for membership of the EU, in Countries which form the southern and eastern borders with the EU, and in partner Countries in other regions.

The EIB Group has been financing SMEs since 1968, and its Governors – the 27 Finance Ministers of the European Union – made this one of its five operational priorities in 2005. Until 2007, the EIB Group's activities in the SME sector comprised three types of product, which were provided through more than 200 commercial banks and almost 300 private equity funds in the 27 Member States of the European Union:

- EIB credit lines to commercial banks for the medium or long-term financing of physical investment by SMEs: this lending activity had an annual volume of almost 5 bln euros and reached some 50,000 SMEs per year.
- Venture capital: since 2000, the EIF has invested 4.4 bln euros in 273 private equity funds which acquire stakes in SMEs, primarily innovative small businesses with high growth potential (the Lisbon Strategy). These funds have enabled more than 800,000 SMEs to be started up across the European Union since 2001.
- Bank guarantees, which are also provided by the EIF. 190 operations have been carried out allowing the partial guarantee of 11.6 bln euros of commitments for bank loans to SMEs. By sharing commercial risks with banks in this way, the EIF shifts the risk, thereby reducing the level of reserves that banks have to accumulate under Basel II. This indirect financial lever stimulates the banking market for SMEs.

The EIB is not a solution for all financing problems, but in the current period, in which most Member States have little room for manoeuvre and the banking sector is faced with a major financial crisis, it can provide flexible and effective additional support. It is of great importance, for example, that the Bank is owned by the Member States, but imposes no burdens on either the European Union's budget or the taxpayers of the Member States, because it borrows funds on the markets independently and without the provision of guarantees by the Member States.

It therefore acts to raise significant volumes of financing at a global level for use in the development of the European economy.

Well before the crisis erupted, the EIB carried out a wide-ranging consultation of SME market players, enabling it to develop a new lending product, dubbed EIB loans for SMEs. These new loans, which have been available since October 2008, and are channelled through commercial banks, are simpler, more flexible and more transparent, thus making it possible to reach a greater number of European SMEs.

The EIB continued its crucial contribution to the European SME sector in 2010 with a total of 10 bln euros in support of this important market. It took signatures of EIB loans for SMEs with intermediary banks to some 30 bln euros between 2008 and 2010, thereby meeting the target set in the 2008 European Economic Recovery Plan one year early. More than 160,000 SMEs received finance from the EIB during this three-year period, helping to fill the funding gap for SMEs, which have found it particularly difficult to obtain financing in these challenging economic circumstances.

EIB loans for SMEs can support all types of investment or expenditure necessary to grow a small business. Hence, the investment can be:

- *tangible:* that is, the purchase of plant or real estate. In principle, the purchase of land is ruled out unless it is vital for the investment, while the purchase of agricultural land is totally excluded.
- *intangible:* such as the financing of expenditure directly related to research and development, building up or taking over distribution networks, including in another EU Member State, the filing or acquisition of patents or the costs incurred in the transfer of an enterprise enabling the continuation of economic activity (where the buyer and the enterprise that are for sale are SMEs, and the amount required to finance the transfer does not exceed 1 mln euros), the permanent increase in working capital required to develop an expanding SME.

Moreover:

- EIB loans for SMEs can support investment in any economic sector except arms, gambling, tobacco, activities involving animal testing, activities whose environmental impact cannot be mostly mitigated or offset, sectors that are morally or ethically controversial (such as human cloning) and pure property development. Also excluded are purely financial transactions (such as company takeovers), with the exception of transfers of businesses in the specific circumstances described above.
- EIB loans for SMEs can support investment of any amount ranging from very small projects to investments costing up to 25 mln euros. The maximum amount provided by the EIB may not exceed 12.5 mln euros per loan.
- EIB finance can be obtained for loans of generally between 2 and 12 years. The actual length of the loan will depend on the economic life of the project financed. EIB loans for SMEs are provided throughout the European Union through commercial banks, which are responsible for evaluating each loan application submitted by a SME. For most operations, it is entirely up to the intermediary bank to decide whether or not to grant a loan to the SME. The EIB's participation serves to enhance the financial terms of these loans, and the EIB ensures that the SMEs concerned are clearly informed of these favourable terms. For certain operations, the EIB can also share the risks with the banking partner or provide "mezzanine" finance.

Recently, by the end of the decade the EIB offers the intermediary banks more sophisticated risk-sharing products designed to reach market segments that commercial banks have difficulty penetrating (that is, SMEs for which the risk is considered too great or the security provided is judged insufficient). Three types of measure are developed:

- loans where risks are shared with the banks, with the EIB guaranteeing part of the total risk taken on by the intermediary bank;
- loans where the EIB takes a risk on the beneficiary SME directly, in parallel with the intermediary bank; and
- mezzanine products for high-growth SMEs or "gazelles". Through its subsidiary the European Investment Fund (EIF), the EIB Group provides participating loans, which

are like quasi-equity and enable fast growing small enterprises to increase their ability to obtain bank credit without necessarily having to open up their capital or provide substantial security (the participating loan being converted into shares only if the company is unable to repay the loan).

Finally, the EIB and the European Commission works together to establish a pan-European microcredit fund aimed at very small enterprises. This fund allows loans to be granted, through around 30 European microfinance institutions, for the creation and development of very small enterprises, as well as to provide technical assistance to such companies. This focuses in particular on companies located in sensitive areas.

3.4.2. The European Bank for Reconstruction and Development (EBRD)

The EBRD is an international financial institution established to favour transition to a market economy in Central-Eastern European Countries and of the former Soviet Union, and the privatization process of their economies. In order to fulfill this mandate, the Bank, which operates in close collaboration with national and international financial institutions, acts directly by granting medium to long term loans, and indirectly through financial intermediaries by granting lines of credit and co-financing. The Bank has 62 shareholders, which include all the Member States of the European Union, the Countries in which it operates, the European Commission, and the European Investment Bank.

The EBRD encourages co-financing and direct foreign investments by the public and private sectors, and promotes local SMEs in order to improve productivity and commercial situations. The EBRD works both directly, by granting medium - to long-term loans and taking shareholdings and quasi-shareholdings (but always in the form of a minority and temporary holding) and indirectly, through financial intermediaries, by providing lines of credit, co-financing, shareholdings and investment funds, services for business, and shareholdings in banks.

3.4.3. The European Investment Fund (EIF)

The European Investment Fund was established in 1994 specifically to support SMEs. Its major shareholder is the European Investment Bank, with which the fund forms the "EIB Group". It provides risk capital to SMEs, especially to newly-formed companies and to technology-oriented activities. It also provides guarantees to financial institutions – for example to banks – to guarantee their loans to SMEs. The EIF is not a credit institution, and therefore does not provide loans or subsidies to enterprises, nor does it invest directly in any type of company. It operates through banks and other financial intermediation entities (microfinance institutions, or private equity and venture capital funds), and uses its own funds or funds entrusted to it by the EIB or the European Union.

The Fund operates in the Member States of the European Union, in Croatia and Turkey, and in the three EFTA Countries (Iceland, Liechtenstein and Norway).

3.4.4. Microfinance

The EU supports individuals facing uncertainty in the job market who wish to set up their own micro-enterprise or opt for self-employment through the European Progress Microfinance Facility. A (future) entrepreneur or a micro-enterprise with up to ten employees looking for a micro-loan up to 25,000 euros under Progress Microfinance can obtain further information from a microfinance provider in its own Country participating in the Facility and acting as EIF intermediary under Progress Microfinance. The EIF does not provide direct financing to entrepreneurs. The financial intermediary informs an applicant about specific requirements and procedures, which vary according to each local market. Applicants can benefit from more than one micro-credit under the initiative.

Financial intermediaries can be non-bank microfinance institutions (MFIs), micro-banks, commercial banks, guarantee institutions and comm-unity development finance institutions operating in all 27 Member States of the EU. The Facility is not made available to intermediaries in EU candidate Countries or potential candidates.

3.4.5. JEREMIE - Joint European Resources for Micro to Medium Enterprises

The JEREMIE initiative developed in cooperation with the European Commission offers EU Member States, through their national or regional Managing Authorities, the opportunity to use part of their EU Structural Funds to finance small and medium-sized enterprises (SMEs) by means of equity, loans or guarantees, through a revolving Holding Fund acting as an umbrella fund. The JEREMIE Holding Fund can provide selected financial intermediaries with SME-focused financial instruments, including guarantees, co-guarantees and counter-guarantees, equity guarantees, (micro) loans, export-credit insurance, securitization, venture capital, Business Angel Matching Funds and investments in Technology Transfer funds.

The main features of JEREMIE are:

- *flexibility:* contributions from the Operational Programmes to the JEREMIE Holding Fund are eligible for interim up-front payments by EU Structural Funds, giving Managing Authorities more flexibility in allocating these resources. Structural Fund contributions to the Holding Funds must be invested in SMEs by 2015;
- *benefits of a portfolio approach:* the Holding Fund is able to re-allocate the resources to one or more financial products in a flexible way, depending on the actual demand over time. The umbrella fund approach allows a diversification of risks and expected returns due to financial products with different default rates, as well as active cash flow management to allow for a swift response to changing market requirements.
- *recycling of funds:* the Holding Fund is of a revolving nature, receiving repayments from the financial intermediaries for further investments in the SME sector. This makes SME support via EU Structural Funds sustainable, unlike the pure grant approach.
- *leverage:* a significant implied advantage of JEREMIE is its potential ability to engage the financial sector either at the Holding Fund level, with additional capital from financial institutions, or at the level of financial instruments, through co-financing, for example, in both cases potentially in cooperation with the EIB.

The Holding Fund develops an Investment Strategy prior to the signature of a JEREMIE Funding Agreement between the EIF and a national/regional government of a EU Member State. The national/regional counterpart transfers the funds allocated to JEREMIE to a JEREMIE bank account, a call for expression of interest is launched and Financial Intermediaries are selected on the basis of a due diligence process. The decision is taken with the involvement of the Investment Board. The EIF's expertise as a Holding Fund manager can be of particular added value in the lesser-developed regions/Member States, where there is a need for capacity-building initiatives and transfer of know-how between local institutions and EIF. In those regions where JEREMIE is managed by another body, EIF can also be involved as an adviser for a wide range of services, such as cash flow management of Holding Funds, structure of Holding Funds, product design in line with European Regional Development Fund regulation, reporting and monitoring, corporate governance due diligence/second opinions, and setting-up of financial vehicles.

3.4.6. JASMINE - Joint Action to Support Micro-Finance Institutions in Europe

JASMINE is a joint initiative of the Commission, the European Investment Bank (EIB) and European Investment Fund (EIF)[74]. It was launched to:

i) *enhance the capacity of non-bank micro-credit providers/micro-finance institutions* (MFIs) in various fields such as institutional governance, information systems, risk management and strategic planning; and
ii) *help them become sustainable and viable operators* in the micro-credit market.

The European Investment Fund (EIF) was appointed by the Commission to set up and manage the JASMINE Technical Assistance Facility, which provides support to micro-credit providers. Provision of technical assistance through JASMINE started in 2010.

4. Incentives for SMEs in the Main European Countries

4.1. Constraints to SME Financing

SMEs, whilst frequently more flexible in investment decisions than larger companies and quicker in response to market inputs, are commonly presumed to be more exposed to many forms of financing constraints. For example, some of them may in practice be excluded from access to public financial markets, while others run ventures and/or are in life-cycle phases such that they are denied access to bank credit (a situation defined as “non-bankability”).

In fact, it is self evident, and widely discussed in the literature, that credit markets are characterized by a very high degree of information asymmetry. Specifically, the financier[75] cannot properly assess *ex ante* the quality and riskiness of the project to be financed. In credit markets, these information asymmetries cause the well-known problems of adverse selection and moral hazard. A significant body of literature, since the ground-breaking 1981 paper by Stiglitz and Weiss, has demonstrated how credit market features may cause the profit function of lenders not to be monotonically increasing in the interest rate, or even in collateral requirements. This amounts to saying that, above a certain threshold of interest rate or collateral (or both), only the entrepreneurs that are more at risk apply for credit, so that banks experience higher default rates with higher levels of the two variables. In particular, Stiglitz

[74] JASMINE is a pilot initiative launched in 2008 by the European Commission, the EIB Group and the European Parliament. The initiative is the operational outcome of an official communication released by the European Commission in November 2007 addressing “A European initiative for the development of micro-credit in support of growth”.

[75] Meaning both as lender and equity subscriber.

and Weiss (1981) showed how an increase in interest rates could make bank profits decrease due to the adverse selection effect; they argued further that even "increasing collateral requirements (would), under plausible conditions, lower the bank's return". The results of these findings are that credit rationing occurs, and neither the interest rate nor the collateral are effective instruments for equating the supply of loanable funds with the demand for them.

Empirical evidence shows that this is particularly true for:

- smaller enterprises, as financial constraints are inversely related to firm size (Bagella, Becchetti and Caggese, 2001);
- start-ups, innovation-oriented and young firms, because of non-bankability due to an uncertain track record and the limited availability of tangible assets (Berger and Udell, 1998).

The data reported in Table 3 show that the afore-mentioned problems are actual for Euro Area SMEs, confirming that these are subject to the impact of imperfections in financial markets more than others.

Cressy and Oloffson (1997) extensively discussed the theory and past evidence on financing constraints for European SMEs. In 2011, indeed, approximately one-quarter of SMEs felt uncomfortable with bank credit and one-fifth with equity financing, while the corresponding data for large companies were both less than one-seventh.

This confirms that there is an imbalance between the demand and the supply of finance, especially for SMEs. Nonetheless - as reported in a Finnish report (Murray et al., 2009) - "determining the existence, magnitude and materiality of such a gap and finding the appropriate form and magnitude of government intervention to address the gap in a given region or at a given point in time are less clear".

The consequence of financing constraints is that not all investment projects with a positive Net Present Value (NPV) - once the weighted average cost of funds has been taken into consideration[76] - can find financing This, in the end, makes the issue relevant to our discussion, because, independently of the instruments/market or the origin, it results in under-investment and – as a consequence – lower economic activity.

Table 3. Confidence in talking of- and obtaining finance of Euro Area enterprises (April to September 2011)

	with banks (%)					with equity investors (%)				
Enterprise (respondent)	*Positive*	*Negative*	*Not applicable*	*Doesn't know*	Total *Positive*	*Negative*	*Not applicable*	*Doesn't know*	*Positive*	Total *Negative*
Large	76	14	9	1	100	26	13	60	2	100
SMEs	63	27	9	2	100	14	21	63	1	100
..Micro	53	34	11	2	100	11	24	64	1	100
..Small	68	23	8	1	100	16	20	63	1	100
..Medium	75	16	7	1	100	19	18	62	2	100
Total	67	23	9	2	100	18	19	62	1	100

Source: ECB (2011).

[76] This is the so-called WACC (Weighted Average Cost of Capital), which takes into consideration the return expected by all lenders to the firm, and the specific financing mix (Damodaran, 2010). Hence, the NPV of the project already considers the cost of funding, where the latter is not a good indicator of rationing.

Table 4. Gross investments in tangible[77] assets per employee (relative to large companies; large = 100 in each year)

Enterprise	*2005*	*2006*	*2007*	*2008*	*2009*	*2010*	*2011*
micro	33	34	32	32	31	31	31
small	37	42	43	43	38	38	38
medium	65	71	71	72	66	66	65
total SMEs	39	42	41	41	38	38	38
large	100	100	100	100	100	100	100
Total	50	53	52	52	50	49	49

Source: our calculations based on EUROSTAT and ISTAT data.

At this regard, Table 4 reports an estimate of the investment gap between larger companies and SMEs for Italy, a Country where the influence of SMEs in national GDP is predominant.

4.2. Incentives for SMEs

SMEs represent 99.8% of EU enterprises, and contribute 66.9% to private sector employment and 58.4% to value added (Ecorys, 2010). It is natural for public authorities to try to solve all problems relative to their investment capability. Financing is one part of the problem. In fact there are two separate issues: on the one hand the availability of different (and appropriate) sources of funds to make projected investments possible; on the other the expected return on the investments to be implemented. Incentives can be directed to both sides, separately or jointly and often with varying mechanisms and levels of effectiveness. This point has already been treated in the Introduction, with a discussion of incentives in general, and an attempt to provide a taxonomy of the different types. Here we wish to address the nature of these measures in terms of their economic impact on targeted firms and, therefore, of their effectiveness. In this regard, we can divide the possible measures to foster investment activity into two broad categories:

1. those directed at increasing the expected return, operating:
   a. mainly on the asset side, in terms of tax credits aimed at heightening expected gross returns on capital invested,
   b. sometimes, and applied especially in the past, on the liability side, in terms of the various forms of contribution to interest expenses;[78]
2. those directed at the availability of funds, in terms of:
   a. capital grants;
   b. facilitation of access to bank credit, for example guarantee schemes aimed at mitigating the credit risk incurred by lenders;

[77] Czarnitzki (2006), for example, estimates the same problem for German SMEs with regard to R&D (intangible assets).

[78] Policies of this kind were particularly pernicious in Italy, where debt was heavily facilitated by various means, and encouraged throughout the entire post-World War II period until the early nineties, and produced a massive distortion of SMEs' financial structure towards high leverage ratios.

c. incentives to the operations of professional equity investors, such as venture capital funds or networks of business angels.

The empirical evidence on the effectiveness of these different types is neither extensive nor, possibly, conclusive. Nonetheless, one can state that measures of the former kind, which exert their effects on the expected rate of return, are rarely successful in extending the overall scale of investments implemented by enterprises. In fact, at best they can encourage anticipation or a different timing of investment projects; that is to say, the effects of the incentives consist only in a time-substitution (Adda and Cooper, 2000). In the case of type-1b measures, moreover, there could be a mismatch between the duration of the facilitation and the investment; or, which is the same thing, there may be uncertainty about its overall effects on the investment's expected rate of return.

Cannari, d'Aurizio and De Blasio (2006), for example, in their analysis of state interventions in Italy in terms of both tax credits and capital grants, report that the triggered additional investments were estimated to be approximately zero as compared to employed resources.

Among the type-2 measures – which are directed at the availability of funds – capital grants have been heavily criticized, and are commonly considered to be ineffective. The main reason for this is that this measure, although it can obviously facilitate fundraising by enterprises, it does not provide proper incentives for maximization of returns, and cannot be associated with effective monitoring mechanisms.

The argument is different in the case of facilitations of access to bank credit. Here it is the bank, as with any other lending activity, that deploys its resources in order to ensure professional monitoring of its debtors. When public policy intervenes in the credit market through guarantee schemes, there is evidence of a mitigation in constraints to SME financing (OECD, 2007).

With regard to incentives for the operations of professional equity investors, many theoretical and empirical studies have carefully studied the fact that there may be a so-called "equity gap" in the market[79]: the possibility that certain entrepreneurs seeking risk finance are rationed.

It should also be considered that equity investments are especially needed in all those situations where banks are not well disposed to credit financing, as is the case where risk and uncertainty are very high. This occurs, for example, with firms in innovative sectors, or ones that need to afford asset-side restructuring, or recently-established ones.[80]

Many different tools for stimulating the supply of risk capital have been identified and employed, above all government equity investments, grants, loans, forms of tax relief, and guarantees. The idea is that, once access to equity financing is provided, not only is investment made possible, but the chances of success are also maximized, considering the focus on investor returns and the very pro-active monitoring tools that they usually implement. Moreover, equity investors, through their screening of investment opportunities, may act as an appropriate filter for the selection of the best possible targets for interventions, in terms of the probability of success in the longer term.

---

[79] This gap may be between the informal financing options and the established private equity market: for example, Harding (2000) with reference to the U.K.

[80] Innovative and recently-established firms are among the greater candidates to spurr innovation (Kortum and Lerner, 1998; OECD, 2006).

In the end, the latter type of intervention seems to be the closest to the relevant managerial issues of both enterprises and lenders, despite the fact that the supply side is only a part of the problem. A strong and aware demand side for risk capital is also crucial for a proper equity market for non-public companies. Harding (2002) significantly points to a possible "knowledge gap": "*...those supplying venture capital do not know about the good, investable projects across all economic sectors. Those demanding venture capital are not aware that venture capital is a suitable means of growth finance or do not want to give up some control of business to allow it to grow.*" It has therefore been claimed that policies also need to be directed towards stimulating the SME sector in general, and promoting a widespread entrepreneurial culture (Queen, 2002; McGlue, 2002). Indeed, Black and Gilson (1998) have argued that a well-developed stock market plays a major role in fostering equity investments.[81]

A rationale for this has been identified in something beyond the simple divestment opportunity for professional investors and which has to do with the attractiveness of equity finance on the demand side (entrepreneurs). In fact, with a concrete "go public" option, there could be the opportunity for implicit contract over control,[82] so that an entrepreneur can reacquire it by procuring an investor exit through an IPO.

4.3. Incentive Programmes for SMEs in the EU

Public administrations usually acknowledge the need for intervention in order to stimulate SME investments and operations, even though – as discussed earlier – there is little evidence of the effectiveness of most of the instruments available. Individual Countries adopt their own domestic measures, if they are permitted under EU state aid rules, in order to stimulate SME investments on the asset side.[83] Tables 4 and 5 below illustrate data on the aid disbursed in EU Countries. First of all, it is evident that grants and tax exemptions are still the main instruments used by administrations. The figures are in line with the following:

- a wide variety of resources are employed in different Countries, but in the main ones (Germany, Italy, France, the UK, and Spain) the total amount is around 0.5% of GDP;
- the percentage of total resources directed to SMEs is very low at an aggregate level (4.2%), but is considerable in Italy (18%, the highest level among all the Countries) and negligible in Germany (2%);
- the main objective of interventions is investments and R&D expenditure (17.9% of the total at European Union level), especially in some Countries, where it accounts for more than twenty-five per cent of the total, such as Spain - among the larger Countries -;
- aid to risk capital is relevant only in the UK (15% of total).

[81] Similarly, Michelacci and Suarez (1998).

[82] As has been suggested, entrepreneurs' fears of losing control may be one of the main factors hindering greater demand for VC financing.

[83] These are measures such as those listed in point 1 in the previous subsection, which are directed at increasing expected returns.

Table 5. Expenditure by aid instrument

Aid instrument (year 2010)	In billion €	% of total aid to industry
Grants	33.5	54.9
Soft loans	1.674	2.7
Guarantees	1.575	2.5
Equity participation	0.6	1.0
Tax exemptions and deferrals	23.7	39.0

Source: our calculations on DG Competition data.

On the other hand, the focus of EU policy, which today is developed under the umbrella of the Small Business Act (SBA) adopted in 2008,[84] is mainly on facilitating the availability of funds. In this regard, the European Commission and the Member States have implemented a comprehensive system of policies, with a wide range of instruments, in order to support SMEs. In particular, improving access to finance is the first of the twelve projects of the Single Market to be implemented by 2012, and the Innovation Union Flagship Initiative Communication devotes two chapters to the finance issue, explicitly mentioning SMEs. The European Commission has also established a permanent SME Finance Forum to monitor the market situation and to encourage new approaches to improving access to finance for SMEs. More in detail, the wide variety of initiatives is directed both at risk capital availability and to access to credit.

4.3.1. Risk Capital

In the context of EU policy, risk capital comprises funds made available to a company during its early-growth stages (start-up and development), and covers three types of financing: informal investment by business angels,[85] venture capital, and stock markets specialising in SMEs and high-growth companies. For the first source "The Commission's policy is to identify and spread good practices that can help improve the conditions for business angel investment"[86]. Then, the availability of early-stage investment is facilitated under the Competitiveness and Innovation Framework Programme (CIP). This includes a set of financial instruments – implemented through financial institutions in the Member States, mainly venture capital funds – that support equity investment. The 2007-2013 Competitiveness and Innovation Framework Programme (CIP) has a budget of over 1bln euros to facilitate access to finance[87] for SMEs where market gaps have been identified. Each euro spent leverages 6 euros on average on risk capital or 50 euros on bank loans, which means it could generate some 30 bln euros in new finance for SMEs from financial institutions. The CIP's financial instruments are implemented for the Commission by the European Investment Fund (EIF) on a trust basis. The EIF is a public-private partnership financed by the European Investment Bank (61.2%), the European Commission (30%) and financial institutions (8.8%), and it is a specialist provider of SME risk finance across Europe and a major player in the European venture capital market. Its investment strategy has

[84] See subsection 3.2.
[85] The term usually refers to private individual high net wealth investors.
[86] See http://ec.europa.eu/enterprise/policies/finance/risk-capital/business-angels/index_en.htm.
[87] The budget refers to both equity and access to credit interventions.

historically focused on early stage VC funds and the promotion of European technology, but it is currently seeking a more diversified portfolio. Since 2007, the European Investment Fund has operated a High Growth and Innovative SME Facility (GIF) providing risk capital for innovative SMEs. The initiative covers different needs, depending on the stage of development of the small business. It provides:

1. risk capital for innovative SMEs in their early stages (GIF1): the EIF usually co-invests 10% to 25% of the total equity with a private intermediary venture capital fund[88], or up to 50% in specific cases;
2. risk capital for SMEs with high growth potential in their expansion phase (GIF2): the EIF co-invests 7.5% to 15% of the total equity with a private intermediary venture capital fund or, exceptionally, up to 50%.

Table 6. Non-crisis State aid disbursed in the EU in year 2010

Countries	In € billion (2010)	As % of GDP	Difference when compared to previous year (by % of GDP)	Trend 2008-10 (in % of GDP)	% of total to SMEs	% of total to Inv. and R&D	% of total to risk capital	% of total to exports and internationalization
EU-27	61.0	0.50	-0.01	0.49	4,2	17.9	1.3	0.4
Belgium	1.8	0.52	0.04	0.45	12	43	0	0
Bulgaria	0.0	0.04	-0.05	0.06	4	16	0	0
Czech Republic	0.9	0.65	0.13	0.66	3	27	0	0
Denmark	2.0	0.83	-0.09	0.82	0	12	0	0
Germany	14.7	0.59	-0.03	0.59	2	21	0	0
Estonia	0.0	0.10	0.02	0.09	7	4	3	0
Ireland	0.9	0.56	0.10	0.51	4	27	3	0
Greece	1.8	0.78	0.02	0.69	10	0	1	0
Spain	4.3	0.41	-0.05	0.43	3	26	0	0
France	12.6	0.65	0.03	0.60	5	14	0	0
Italy	3.3	0.21	-0.09	0.27	18	17	0	6
Cyprus	0.1	0.51	0.13	0.45	5	2	0	0
Latvia	0.1	0.40	0.28	0.24	1	6	0	0
Lithuania	0.1	0.29	0.03	0.23	8	14	0	0
Luxembourg	0.1	0.18	-0.07	0.19	14	57	0	0
Hungary	1.9	1.94	0.56	1.72	1	5	0	0
Malta	0.1	1.12	-0.48	1.47	0	0	0	0
Netherlands	1.9	0.32	0.01	0.29	3	37	0	0
Austria	2.1	0.72	0.08	0.65	4	25	1	0
Poland	2.5	0.72	-0.01	0.72	0	3	1	0
Portugal	1.5	0.90	-0.06	0.92	1	4	0	0
Romania	0.2	0.17	0.03	0.17	2	16	0	0
Slovenia	0.3	0.89	0.10	0.71	0	30	0	0
Slovakia	0.2	0.37	0.01	0.39	0	7	0	0
Finland	0.8	0.43	-0.04	0.44	4	31	2	5
Sweden	2.6	0.76	-0.06	0.80	0	4	0	0
UK	4.1	0.24	0	0.23	5	20	15	0

Source: our elaborations on data from DG Competition.

[88] It should be noted that the intermediary venture capital funds make investment decisions based on normal commercial and financial criteria.

Certain actions are conceived in order to develop growth[89] stock markets and make them easily accessible by firms so that they can be listed easily. For example, the Commission is seeking "to facilitate cross-border operations in financial exchanges, remove obstacles to the use of competing clearing and settlement systems, and apply common rules to trading".[90]

The policy idea, which is well grounded in financial theory and literature, is that initial public offerings (IPO) of stock provide a natural exit route for venture capital investors, and give valuation guidance for sales to other companies. Without liquid exit markets for venture capital - that would also have a critical mass of advisory services around them - venture capital funds will always face difficulties.

4.3.2. Access to Credit

Firstly, the European Commission actively tries to facilitate the efficient functioning of credit markets: for example, it closely monitors the markets for business loans and provides tools and information to help businesses, such as the guide for SMEs on how to work with banks (July 2005). Moreover, since 1998 the European Union has used a set of financial instruments (mainly guarantees and microcredit[91]) to increase the volume of credit available for small businesses.

In the area of guarantees, for the period 2007-13 there are instruments that are also part of the Competitiveness and Innovation Framework Programme (CIP). In particular, the SME guarantee facility (SMEG), which is administered by the EIF, provides loan guarantees to encourage banks to make more debt finance available to SMEs, including microcredit and mezzanine finance, by reducing the banks' exposure to risk. The SMEG provides financial intermediaries financing SMEs through loans, or mezzanine finance, or microcredit or equity with:

i) co-guarantees,
ii) counter-guarantees,
iii) direct guarantees,
iv) securitisation – guarantees, to support the mobilisation by financial intermediaries of loans to SMEs.

To date, the SMEG programme has been very effective in reducing the financing gap for many SMEs.

Table 7 shows the distribution and size of loans backed by SMEG guarantees across participating Countries. Some 90,000 enterprises have benefited from the guarantee, especially in France, Italy, Germany and Spain.

As for microcredit, the EU works actively to help Member States and regions improve their provision of microcredit to entrepreneurs through policy learning and by spreading good practice. For example, in 2007, the Commission launched an initiative to seek to improve the availability of microloans for small businesses and for socially excluded people wishing to become self-employed. The scheme operates on four fronts:

---

[89] These markets provide listing opportunities with simplified requirements for firms that cannot or do not want to be listed on the main stock markets.

[90] See http://ec.europa.eu/enterprise/policies/finance/risk-capital/index_en.htm.

[91] Defined as loans of amounts up to 25,000 euros.

- improving legal and institutional conditions for businesses in EU Countries, for example by relaxing interest rate caps, providing access to borrower databases and tax incentives, and adapting banking regulation;
- promoting a positive view of entrepreneurship so as to encourage people to become self-employed;
- spreading best practices for microfinance institutions;
- topping up financial capital for micro-credit institutions, with a focus on new and non-banking institutions.

Apart from the SMEG under the EIF, there are many other funding facilitations for microcredit, such as in the JEREMIE and JASMINE programmes,[92] or under the European Progress Microfinance Facility (EPMF), that directly provide microcredit to small businesses and to people who have lost their jobs and wish to start their own small businesses.

It is worth noting that the European Investment Bank[93] (EIB) and the European Investment Fund (EIF) are now also developing a mezzanine financing instrument. This is suited to providing support to businesses in various phases of their life-cycles – such as the growth or transfer of the business – in the form of mezzanine finance, which generally speaking combines the features of loans and equity in the risk-return profile in the same instrument.

Furthermore, and finally, the European programme known as "Eurostars" should be mentioned in regard to EU policies targeting SMEs. It has the objective of supporting R&D-performing entrepreneurs by funding their research activities.

Table 7. Support provided to SMEs through the SMEG loan window (at March 2011)

Countries	number of loans	value of loans (average loan size)	number of SMEs reached
Austria	430	22,916	414
Belgium	2,352	44,981	2,271
Bulgaria	230	33,954	225
France	36,834	31,702	32,222
Germany	15,563	29,785	14,567
Hungary	19	354,147	15
Italy	17,262	90,936	16,011
Latvia	6	152,625	6
Poland	0	0	0
Slovenia	709	255,448	664
Spain	16,102	142,012	13,908
EU	89,507	n.a.	80,303

Source: our calculations on data from "Center for Strategy and Evaluation Services (2011).

[92] See subsection 3.4.
[93] See subsection 3.2.

Eurostars will run initially from 2007 to 2013, and combines the domestic R&D programmes of 33 Countries, which have committed 300 mln euros, the European Commission's Framework Programme 7, with up to 100 mln euros, and the Eureka Secretariat (for co-ordination).

In addition to policies implemented directly by EU bodies and funds, many Member States have adopted policy measures in accordance with Principle 6[94] of the SBA. In particular, at a national level:

- many guarantee schemes have been implemented (for example in Italy, Germany, France, Spain, and the UK);
- microcredit co-funding has been supported (for example, in Germany, Sweden, and Austria);
- risk capital availability increase has been targeted (Germany, UK, Ireland, Sweden, etc.).

5. The Recent Evolution of Italian Policies for Financial Incentives to SMEs

5.1. The Implementation and Revision of the Small Business Act

As a part of the implementation of the priority indications contained in the SBA Directive, numerous interventions were carried out in the course of 2010.[95] These may be grouped into three major categories.

Among the interventions with the purpose of simplifying procedures and improving the relations between the public administration and enterprises, mention should be made of – besides the entry into force of the "Comunicazione Unica" (Single Communication) (on 1 April 2010) and the "Segnalazione certificata di inizio attività" (SCIA) (Certified Report of Commencement of Activity) – the regulations of the "Agenzie per le imprese" (Enterprise Agency) and the "Sportello Unico per le Attività Produttive" (SUAP) (One-Stop Shop for Productive Activities) (in Presidential Decrees nos. 159 and 160 dated 7/9/2010 respectively).

The second category is that of interventions in support of innovation, including the following:

- the "*Progetti di Innovazione Industriale (PII)*" (Industrial Innovation Projects), which have, by means of three calls for funding applications (Sustainable Mobility, Energy Efficiency, and Made in Italy) facilitated over 230 projects submitted by 1745 enterprises, of which 1268 were SMEs and 500 research and university bodies, for a total of subsidized investments equal to almost 2.2 bln euros, and for a total amount of subsidies granted equal to around 850 mln euros;

[94] This states as its objective a reduction of the financing gap for SMEs.

[95] For a complete review, see MI.S.E. (2011).

- the "*Fondo Rotativo per il Sostegno alle Imprese (FRI)*" (the Rotating Fund for Support to Enterprises), which has been bolstered by an allocation of 785 mln euros for technological and industrial innovation agreements, pursuant to Law no. 46/1982 (Technological Innovation Fund);
- the issue of three ministerial decrees for the Regions which fall within the convergence objective of EU cohesion policy, with the purpose of promoting investments in the renewable energy sector for innovative enterprises;[96]
- and finally, the regulation of the network agreement[97] as a measure to support the aggregation of enterprises in order to improve their innovative and competitive capacities on markets.

The third category includes interventions aimed at facilitating access to credit. They are:

- the augmenting of the "*Fondo Centrale di Garanzia*" (Central Guarantee Fund) for SMEs, created by Law no. 662/1996, with 2 bln euros for the five-year period 2008-2012; the raising of the maximum guaranteed amount from 500,000 euros to 1.5 mln euros; extension of the scheme to the crafts, transport, and cooperative sectors (in less than two years, at the end of 2010, over 75,000 applications had been received, with 15 bln euros of investment guaranteed);
- an extension of the moratorium on SME debts to 31 July 2011;
- the creation in March 2010 of the "*Fondo Nazionale d'Investimento*" (National Investment Fund), a measure to support the capitalization processes of Italian SMEs, with a budget of 1.2 bln euros. The Fund is intended above all to assist companies with turnovers between 10 and 100 mln euros, and with plans for foreign growth; 50% of the resources are used to access companies' capital, while the remaining 50% is to provide resources to private equity funds which invest in SMEs;[98]
- execution of the agreement with the "*Associazione Italiana di Private Equity e Venture Capital (AIFI)*" (Italian Association of Private Equity and Venture Capital), which has the purpose of spreading venture capital among high-level small enterprises.

In February 2011, the European Commission summarized the implementation of the SBA, and created a revised document, with a significant contribution from Italy, in which indications were provided for the future. In short, the principal actions that the Member States were invited to take were the following[99]:

---

[96] The purpose of the first decree was the implementation of interventions on the Interregional Operating Programme POIn "Renewable Energy and Energy Saving" 2007-2013 (with a contribution of 300 mln euros. The second was for projects for the industrialization of the results of advanced research or experimental development programmes, pursuant to the 2007-2013 Research and Competitiveness National Operating Programme (PON) (100 mln euros). The third was for investment programmes aimed at pursuing specific innovation, competitive improvement, and environmental protection objectives, pursuant to the 2007-2013 Research and Competitiveness National Operating Programme (PON) (100 mln euros).

[97] For a more detailed analysis of this measure, see the following subsection.

[98] For a more detailed description of the FII, see subsection 3.6.

[99] Some of these actions had already been undertaken in Italy.

- implementation of the 'Think Small First' principle, both from the legislative point of view and from that of the administrative procedures that regulate enterprises, by introducing a single correspondent and simplifying the obligations arising out of them;[100]
- facilitation of access by SMEs to structural funds, especially in support of the creation of networks of enterprises, with priority in the environmental and energy sectors by SMEs;[101]
- ensuring that double taxation and inconsistencies in tax treatment do not hinder cross-border investments in venture capital;
- considering the creation of one-stop shops where SMEs can apply for European, national, and local contributions;[102]
- facilitation of access by SMEs to public tenders (for example, by dividing contracts into lots, and avoiding selection criteria that penalize SMEs);
- utilization of a single, centralized website for publication of tender opportunities, which includes the function of "finding a business partner" for a joint offer, sub-contracting or other commercial opportunities;
- further development of support for the transfer of enterprises, in particular of auditing businesses, and the creation and promotion of an on-line data-base and platforms, thereby facilitating meetings between entrepreneurs and possible purchasers.

Finally, the Italian request that the institution of "regional SBAs" be promoted to take account of territorial specificities existing not only in Italy but throughout Europe was accepted.

5.2. Tax Subsidies Connected with the "Network Contract"

Law no. 5/2009 (at Article 3, paragraph 4-ter) regulates network agreements as a form of cooperation among enterprises in order to increase their competitiveness on the market, since this new legislation is aimed above all at small and medium enterprises.

More specifically, a network is a multilateral associative agreement by which two or more enterprises undertake to establish a common organization for the purpose of jointly carrying out an economic activity that falls within the corporate purpose of each enterprise, in order to increase their innovative capacity and competitiveness on the market (Moscatelli, 2011).

Cooperation among enterprises is a fairly widespread phenomenon in the international field, where commercial practice shows an increasingly favourable attitude towards the

[100] Italy is noteworthy for having promoted numerous initiatives of this nature: for example, in 2010, there were almost 3,300 one-stop shops at town council offices, of which 25% were already automated.

[101] Here, Italy has suggested the introduction of a "European Network Agreement" to promote relations among EU SMEs and to support the internationalization process.

[102] In this regard, it should be clarified that the creation of these one-stop shops in Italy was a part of the then-existing process of reorganization and rationalization of bodies operating in support of Italian enterprises abroad.

execution of joint ventures agreements; in the future, therefore, if not immediately, the network agreement format may find appropriate application also in Italy.[103]

Notwithstanding the significance of the law, there is currently only one tax subsidy in favour of network agreements: the one introduced by Article 42 of legislative Decree no. 78/2010, which takes the form of a suspension of tax (income tax and IRAP (the Regional Tax on Productive Activities) for investments carried out by enterprises which apply the network programme.

It should be emphasised that the decision to offer tax concessions has been endorsed by the European Commission in that it is consistent with the lack of a subjective nature of the network, because the special treatment is given to individual enterprises in relation to their individual choices, not to the network as a whole.

The elements required for application of the tax break are participation in the agreement, confirmation of the programme, and the destination of profits, as well as the temporal aspect. Network investments must, in fact, be completed by the end of the tax year following the one in which the resolution regarding the profits provision was approved.

Finally, the tax concession must also have a time limit, inasmuch as it applies to profits set aside from the tax period in course at the end of 2010 until the tax period in course at the end of 2012, and may not exceed the ceiling of 20 mln euros for 2011, and 14 mln euros for each of 2012 and 2103 (Tassani, 2011).

Brancati (2011) has criticised this legislation on the grounds that in addition to giving a very complex definition of the beneficiaries, the implementing provisions are very difficult to define, so that their actual application will not be straightforward. If this arduous task is successful, there remains the fundamental question of the total financial endowments, which, in theory, could be allocated to at least a third of the Italian productive system (and therefore to a huge number of enterprises); since, as we have seen, this concerns 14 mln euros a year (20 in the first year), they are modest indeed. The Regions will add further resources, but these amounts certainly do not seem to be adequate for the objectives and expectations that are being created, where the management mechanisms, which are fiscal in nature, and therefore highly attractive to operators, would favour very intense use of the instrument. "As things stand, we are speaking of a board game which lacks numerous essential elements with which to be able to properly define a policy with some hope that it might be effective" (p. 1).

5.3. The "Cassa Depositi E Prestiti" (Deposit and Loans Bank)

Within the sphere of activities in support of the economy, the "Cassa Depositi e Prestiti" (CDP) (Deposit and Loans Bank) plays a key role in supporting enterprises, above all small and medium-sized ones. In performing this function, the Bank acts with the support and collaboration of the Italian banking system. In particular, the cooperation of the "*Associazione Bancaria Italiana (ABI)*" (the Italian Banking Association) has enabled the CDP to use special agreements to establish specific funds whose purpose is to favour access to credit by SMEs.

---

[103] It should be noted that the legislative evolution that led to the introduction of the network agreement into the Italian legal system has been rather lengthy and troubled, notwithstanding repeated solicitations both from the business world and from legal doctrine that regulations be defined with the aim of governing the various typologies of "collaboration" and "integration" among enterprises. See Iamiceli (2009).

The SME fund, which has applied since the second half of 2009, guarantees financial coverage of investment costs or working capital increase requirements in the business sector.

The "*Plafond Accordo PMI*" (the SME fund Agreement), created in March 2011 with a bln euros, permits SMEs under certain conditions to secure an extension of the term of loans contracted with the banking system, while maintaining the originally agreed rate of interest.

The CDP also participates, together with certain credit institutions (Monte dei Paschi di Siena, Intesa San Paolo, and Unicredit), as well as other investors, in the equity fund for small and medium-sized enterprises known as FII – the "*Fondo Italiano di Investimento*" (the Italian Investment Fund), which is sponsored by the Ministry of the Economy and Finance, ABI, and Confindustria (the Italian Industrialists Association). The dual objective of the FII is to facilitate capitalization processes and incentivize aggregation among enterprises, in order to generate a significant nucleus of "national champions" in the medium term which is sufficiently well capitalized and, therefore, is able to face the challenges of international competitiveness.

5.3.1. The SME Fund

Within the area of new schemes to support Italy, the CDP has made available a fund to sustain access to credit by SMEs. It covers the requirements of a market in which, in the face of abundant short-term liquidity, the need to make longer-term provisions is emerging, and this can be guaranteed by a real medium- to long-term operator such as the CDP.

The criteria for assigning the total resources have been defined in three phases, by separate agreements between the CDP and ABI: a first Agreement, signed in May 2009, launched the initiative by regulating the use of an initial *tranche* of resources; the second Agreement, signed in February 2010, served to frame a more detailed structure of the product, with further information regarding the deadlines for the funding offered by the CDP. In addition to providing a further, ten year, deadline, the third Agreement, signed in December 2010, provided the structure with greater stability.

The distribution channel is the banking sector, so that in order to have access to funding, SMEs apply to the banks that are a part of the initiative. The CDP's special-purpose funding must be used by the credit institutions to cover the SMEs' investment costs, or to cover their requirements for increasing their working capital. Contracts between the banks and the enterprises must specify that recourse has been made to CDP funding, as well as the collection costs incurred by the credit institution.

5.3.2. The SME Fund Agreement

In February 2011, the Presidency of the Council of Ministers, the Ministry of the Economy and Finance, ABI, and the other associations representing enterprises signed the "*Accordo per il Credito alle Piccole e Medie Imprese*" (the Agreement for the Provision of Credit to Small and Medium Enterprises) (the SME Agreement). In line with the current framework for the period in question, this document defined certain measures for exiting the so-called "*Avviso Comune*" (Common Notice/) dated 3 August 2009, which – as noted – gave SMEs the opportunity to suspend payment of the share of principal of their existing loan installments with the banking sector (for a period of twelve months).

Among these measures was the opportunity, under certain conditions, for SMEs to secure an extension of the term of the payment period for existing loans with the banking sector which were already beneficiaries of the loan installment suspension pursuant to the Common

Notice for a maximum period equal to 100% of the remaining term, and in any event for a period no longer than two years for unsecured loans, and three years for secured loans. In this context, the banks undertake to maintain the original contractually agreed interest rates unchanged if the loan term extension operations are assisted – *inter alia* – by the CDP fund.

For this purpose, the CDP has made available to the banking sector an amount equal to one bln euros, which is regulated by a new Agreement between the CDP and ABI signed in March 2011.

Finally, the CDP has a share of the FII worth 250 mln euros. This is a closed-end fund which makes medium- to long-term investments in companies with turnovers between 10 and 100 mln euros. There are, therefore, around 15,000 potential candidate enterprises in Italy, excluding those of a smaller size.[104]

5.4. The Italian Investment Fund

As said, the purpose of the FII is to create a greater share of medium-sized companies over the medium term by incentivizing aggregation among smaller enterprises so that they become more competitive in international markets as well.

More specifically, the Fund provides small and medium-sized Italian enterprises with financial support, within the sphere of "expansion capital" operations designed to fund the development of existing enterprises, also through acquisitions, by direct or indirect investments in risk capital.

Also to be noted are replacement operations intended to restructure the shareholding base, where the Fund may take the place of minority shareholders no longer with an interest in pursuing the company's activities, or to management buy-ins and buy-outs, where the aim is to support the acquisition of enterprises hampered by problems relating to generational transfers, and to develop potential aggregations by involving managers inside or outside the enterprise.

The Fund should invest in enterprises with a solid equity position, sufficient revenues (both current and forecast), and above all with significant potential for value creation. Within this investment policy, special attention is paid to investments for small and medium-sized enterprises:

- that have interesting development prospects, both domestically and (above all) internationally;
- that intend to undertake concrete aggregation projects with the aim of strengthening their position in the domestic and/or international market;
- that have the potential to exploit trademarks, patents, or special know-how, or that use advanced and innovative processes and technologies;
- that are family-run, but where there is the prospect of generational or entrepreneurial changes;

[104] The result of this is therefore that small enterprises (that is, those with a turnover of less than ten mln euros), despite the recommendations of the European Commission as far back as 2003 which defined a small enterprise as one with a turnover below this amount, are excluded from the FII. The only means of access to the Fund by small enterprises therefore remains that of creating consortia, so as to overcome this limitation.

- that have entrusted their management to an efficient and highly-experienced and professional management structure.

Investments in the following are excluded, however:

- public companies, as a speculative investment;
- start-up companies;
- real estate enterprises;
- enterprises that provide financial services in general;
- enterprises in a state of crisis;
- enterprises subject to insolvency proceedings, or close to them, and engaged in restructuring plans pursuant to Article 67, paragraph 4, letter d) of decree no. 267 dated 16 March 1942 (Bankruptcy Law), or debt restructuring agreements pursuant to Article 182 bis of the Bankruptcy Law.

Each investment is assessed in light of the Industrial Plan, the development project, and the management's capacity to accomplish it.

The FII's initial funding endowment is 1,2 bln euros, and the first direct investments have been made: the six direct investments in the capital of Italian SMEs examined by the Fund's Board of Directors should have been made by the end of 2011.

5.5. Other Funding Measures for SMEs

Despite what has been illustrated in this section, the concrete opportunities available to Italian SMEs to support their investment projects are not especially numerous.

If one excludes the tax credits provided by Legislative Decree no. 70 of 2011[105], in fact, a number of domestic incentive measures were no longer active towards the end of last year, either because they had not been refunded, or because there was a lack of financial coverage, or because a further implementing provision was being awaited.[106]

The overall reform of incentives that should have been completed in early 2012 and with special access reserved to SMEs, which would be the beneficiaries of at least 60% of automatic and assessed incentives, is also on stand-by. The use of the conditional tense is necessary, if one considers that the delegation to the government – which was originally included in the February 2011 package for re-launching the economy in order to reduce and simplify financial assistance to enterprises – has been subject to a number of delays, and is now included in the Enterprise Law. This law, it should be recalled, in implementation of the

[105] Article 2 of this measure provides for a tax bonus of 50% of wage costs incurred by SMEs for permanently hiring new workers in the Mezzogiorno, while Article 2 bis provides a tax bonus for new investment in disadvantaged areas of the Mezzogiorno.

[106] In brief, the main legal provisions in question at the end of last year were:
Article 3 of the cited legislative Decree 70/2011 (tourist areas), which provides tax, administrative and financial breaks for research activities by tourist enterprises set up in the area;
Article 14 of Law 99/2009 (internationalization of SMEs – Start-up Fund), which provides a direct fund to support SMEs' internationalization projects, whether individually or in groups;
Article 1, paragraph of Law 296/2006 (National Innovation Fund), which provides support for innovative projects based on the economic exploitation of intellectual property rights by SMEs. See Barbieri and Reggio (2011).

2009 delegation established by the Development Law, – provided for the compete reorganization of 100 national and 1,400 regional laws, cancelled around thirty laws and forms of incentive, and simplified the overall framework by introducing three broad categories: automatic incentives (for example, tax bonuses), project funding applications, and procedures for the financing of large projects. Last December's Enterprise Law (Law 180/2011), which was inspired by the European Small Business Act, and designed specifically for SMEs, and which was approved with high praise by all the trade associations representing SMEs, provides (in Article 10) for another postponement of the term for delegation to the government in the area of incentives to enterprises (and their internationalization). The only positive signs for SMEs can be found in the so-called "*Salva Italia*" (Save Italy) Decree (Legislative Decree no. 201/2011), inasmuch as it provides, among the growth measures of a financial nature, for augmenting the "*Fondo di Garanzia*" (Guarantee Fund), thereby opening up new opportunities for these enterprises.

More precisely, Article 3, paragraph 4 provides for a 1.2 bln euro refinancing for the three-year period 2012-2014, the intention being to make the use of the Guarantee Fund even more of an incentive for enterprises. The new laws included in the year-end budget measures lean towards reinforcing an instrument which is considered to be extremely useful for promoting access to credit by smaller enterprises, as emerges from data on the operations which have been completed. The past two years have seen such a large increase in accepted operations that the number has doubled (in 2009, there were 24,600, and in 2010 more than 50,000). Over time, however, this massive recourse to the Fund has also generated dome undesired collateral effects, such as a reduction in the share covered by guarantees from 60% to 50% of the funded amount so that the available funds can cover all requests for guarantees presented by enterprises. With the funds made available by the "*Salva Italia*" Decree, therefore, it will be possible to carry out new developments in favour of enterprises, starting from an increase of the direct guarantee provided by the Fund from 50% to 80%.

Finally, among the expected measures in support of productive investments by SMEs, also to be mentioned is the increase in the guarantee ceiling for each individual enterprise (from the current 1.5 mln euros to 2.5 mln euros), and the reduction, and eventual abolition, of commissions for use of the Fund's guarantee. In practice, the Fund will allow SMEs to secure more credit at a more advantageous cost. While the next reform of incentives is being awaited, we would emphasize that subsidies remain one of the principal themes of economic policy on which the technical-scientific debate is focused, above all on the occasion of public finance budgetary measures, when the Executive is called upon to decide on the allocation of available financial resources. According to the differing attitudes of policy-makers and the economic context of the moment, at certain times incentives are perceived as instruments with which to re-launch the competitiveness of enterprises, at others as tools to recoup delays in the development of various parts of the Country, and at others as a simple "hoard" on which to put the public accounts back in order. While these various interpretations have alternated in recent years, in the past few months the deterioration of the economic crisis has greatly accelerated this alternation. On the one hand, entrepreneurs and trade associations are seeking other forms of assistance in order to face the crisis; on the other, politicians are increasingly divided between supporters and opponents. The latter in particular have adopted the positions expressed by certain economists[107] and authoritative representatives of the banking sector, and claim that incentives are substantially ineffectual and should be replaced by alternative policies for re-launching growth: above all, reductions in taxes and the tax wedge, and the implementation of a new policy for infrastructures.

5.6. An Attempt to Quantify the Resources Employed

Within the context of the implementation of the SBA[108], many interventions have recently been carried out in Italy. In particular, we should recall here:

- the "Industrial Innovation Projects (PII)" directed at facilitating R&D investments, which stimulated 2.2 bln euros of investments with 0.85 bln euros of public resources deployed;
- the "Rotating Fund for Support to Enterprises (FRI)" aimed at facilitating funding to enterprises for investments in R&D in terms of both capital grants and interest rate facilitations, with an additional endowment of 0.79 bln euros;
- the "National Operating Programmes (PON)", co-financed with the EU and conceived in order to stimulate investments in renewable energies, environmental sustainability and industrial innovation, which have been endowed with 0.5 bln euros for 2011;
- the "Central Guarantee Fund for SMEs (FCG)", financed with 2 bln euros for the period between 2008 and 2012, which has guaranteed more than 15 bln euros of investments;
- the "National Investment Fund", launched in 2010 with 1.2 bln euros of capital as a measure to support equity investments in SMEs[109] by means of both direct interventions – for 50% of available resources – and investments by private equity funds as to the remainder;[110]
- the "National Permanent Committee for Microcredit", which supports the development of microcredit and financing microcredit programmes and microfinance institutions;

[107] Here particular reference is made to a recent study by Caprara, Carmignani and D'Ignazio (2010) which provides a statistical framework of the size and composition of special-rate loans granted to Italian enterprises by banks based on public funds. Their interpretation refer to reports to the central Risk Agency (CR), which in the period under review represented an average of 14% of the total of public special-rate loans made to the Italian productive system. Between 1998 and 2007, special-rate loans surveyed by the CR amounted to around 0.3% of GNP, and involved around 27,000 enterprises, mostly consisting of companies. The results which emerged from the estimates confirm that special-rate loans granted on the basis of public funds constitute the predominant form of subsidy in the North and Centre of Italy, while in the *Mezzogiorno*, interest-free loans have historically been the principal source. For the purposes of this work, of principal interest is that the incidence of special-rate funding is higher among larger enterprises, in particular among industries in the strict sense of the word and agricultural enterprises.

[108] For further details, see also the "Rapporto SBA 2010".

[109] Target enterprises are especially those with total sales between 10 mln and 100 mln Euros.

[110] With the aim of spreading the tendency of SMEs to look to outside equity investors, a specific agreement has also been signed between the Ministry for Economic Development (MISE) and AIFI (the Association of Professional Private Equity and Venture Capital Investors).

- the agreement between the "Cassa Depositi e Prestiti (CDP)"[111] and ABI, the National Banking Association, to provide banks with 8 bln euros of funds until 2020 to be used for loans to SMEs.

Table 8. Incentives deployed in Italy (€ millions)

Incentives	2008	2009	2010
Incentives to enterprises[112]	6,559.29	8,925.98	7,651.58
Fiscal incentives	1,157.85	1,538.14	3,635.44
Pension contribution facilitations	1,787.20	1,554.10	688.34
Total	9,504.33	12,018.22	11,975.37
Total as % of GDP	0.60%	0.79%	0.77%

Source: Elaborations from Affuso and Nannariello (2011) and ISTAT national statistics.

Although the foregoing provides a comprehensive overview of the most recent initiatives, the operating framework of public interventions in the area of incentives is very intricate because it results from the stratification of many different instruments and various funding sources (the EU, state administrations, and regions[113]). In order to provide the reader with an idea of the weight and relevance of interventions, we now seek to quantify the amounts of resources deployed – at least by the state, because on the one hand, the EU instruments have already been discussed, and on the other regional policies are so fragmented that is extremely difficult to report on them in a timely manner. To this end, we have included Table 8, which contains data developed by Affuso and Nannariello (2011) directly from the state balance sheet and provides a picture of current efforts by Italy to support enterprises. This picture is incomplete in some respects, because many interventions – even where funded by the state – are represented otherwise in its accounts, and are biased away from the focus on SMEs, because they include resources addressed (a) to state-owned enterprises (such as the railways, for example) and (b) to larger companies (for example, airlines).

The evidence reported shows that the amount of resources dedicated to incentives is approximately 10 bln euros, with a peak of 12 bln euros (0.79% of GDP) in 2010, and a more recent declining trend – as we have also seen above. Incentives, in particular, are divided into three categories: (i) direct transfers of various forms to enterprises; (ii) tax credits and deferrals; and (iii) facilitations in contributions to workers pension schemes[114]. During the period under analysis, the first was the most important type of facilitation: it accounted for 74.3% of the 2009 total. It should be noted, however, that in 2010 there was a significant change of direction, as fiscal incentives grew to a considerable 30.3% of the total.

Overall, the amount of resources employed is not negligible, especially if one considers that other sources (EU and regional) are to be added. The problem is that the instruments are different, and not always comparable in terms of effectiveness.[115]

---

[111] CDP is the central institution of the Italian system for postal savings.

[112] This line represents direct transfers to firms.

[113] These are the regional administrative authorities in Italy.

[114] In Italy, these are compulsory public schemes.

[115] The topic, as discussed above, remains open, and the available evidence is not conclusive. It would certainly be of great interest to investigate the issue of effectiveness further in a European-level comprehensive study taking account of both SBA interventions and national policies.

CONCLUSION

From the foregoing analysis it emerges without a shadow of doubt that the range of economic policy measures for the benefit of micro-, small and medium-sized enterprises (MSMEs) in EU Countries is very broad, in that they attempt to take account of all the possible problems encountered by smaller-sized enterprises.

The principal aim of these measures is to reduce the number of problems that small-scale entrepreneurs are usually obliged to deal with: from under-capitalization of their companies to a limited capacity for initiatives in international markets. Alternatively, they even attempt to eliminate such problems altogether where policy focuses on the growth of the beneficiary enterprises and transforms them into large companies.

Evidently, if the policy is as ambitious as this, it must draw on a large range of instruments consisting of both real incentives (that is, the supply of services) and financial types of incentive, possible integrated in order to improve their effectiveness.

The policy in favour of MSMEs calls for the involvement of various levels of government operating in EU Countries and with competences in the area of industrial policy; in Italy, this even includes local entities such as the provincial administrations, chambers of commerce, and even local municipalities, above all if they are provincial capitals. In fact, the role of the European Commission is almost exclusively that of direction and coordination at a supranational level and the funding, partly directly and partly indirectly, of the measures that it puts into place. The main duties fall to the Member States, which must, with regard to the specific problems complained of by the MSMEs operating on their territories, must take action by applying EU Directives. As a result, the actual results obtained may vary from Country to country as a function of the capacities shown by the beneficiary enterprises themselves, by the domestic public administrations, and by the institutions that have the duty of acting in a manner consistent with and complementary to the European Commission.

With regard to Italy, which is, as we know, particularly interested in MSMEs, and which has for long manifested a political will to intervene on their behalf, the initiatives undertaken at a Community level over the past decades have, obviously, reinforced the tendency shown at a national level, to the extent that according to government sources, Italy is one of the Member States that has implemented the SBA most rapidly and with the greatest conviction.

However, even in the case of Italy, issues arise that, in our opinion, have not been dealt with or resolved adequately. In the first place, the application of the SBA has not been exploited to provide an overall reorganization of the prior, heterogeneous, legislation, above all in the area of financial support, with the consequence that that the oft-announced reform of incentives has not yet been carried out. Furthermore, other provisions have been passed and adopted in recent years, so that the already huge array of instruments operating in various sectors has been further extended (from the expansion and/or renewal of production machinery to technological innovation), and the choices and decisions that small businesses must take have been made even more difficult by the need to comply with the EU's "*de minimis*" regime.

In the second place, the instruments most recently adopted to avert the problem of under-capitalization, and the related requests for real guarantees that banking institutions usually put forward, do not seem to have been used to the extent expected or hoped, either owing to their configuration, which – as is the case with the FII – tends to favour "middle class" enterprises, or because of the grave economic crisis of the past years (since 2008) that has also involved Italian banks, which have resorted to a "credit crunch" in order to avoid the risk of incurring greater debts or even dramatic insolvencies.

Furthermore, a review of the measures adopted most recently in Italy tends to show real signs of a lack of attention and concerns towards MSMEs and their problems on the part of the national Public Administrations, at least when compared with that shown by the European Commission.

We refer to the following matters in particular:

- the fact that measures aimed at MSMEs and larger enterprises, together with the risk that the latter will prevail over the former due to their greater contractual and market power;
- the fact that measures were adopted at a national level – that is, by the central government – which it would have been more appropriate to leave to decentralized administrations, and which must therefore compete with them, with the easily-imagined confusion that this will create among the beneficiaries;
- the almost total lack of availability of official information -a further indication of the merely relative interest in MSMEs - regarding the amount of financial incentives actually granted to smaller enterprises, as evidenced by the failure to publish the annual Ministry report on interventions in support of productive activities, the latest edition of which came out in June 2009.

REFERENCES

Affuso, A., and Nannariello, G. (2011). Incentivi alle imprese: una quantificazione delle risorse. *EyesReg.*, 1 (3), pp.64-69.

Amato, G. (1972). Il governo dell'industria. Bologna: il Mulino.

Ammassari, F., and Marzorati, L. (eds.) (1998). Gli aiuti regionali alle imprese. Temi e problemi, Roma: Autorità Garante della Concorrenza e del Mercato.

Amorth, A. (1931). I contributi pecuniari concessi dallo Stato ad enti pubblici e privati. Urbino: Società tipografico editoriale urbinate.

Antonucci, M. (2003). Brevi cenni sugli aiuti di Stato. *Consiglio di Stato*, 1, pp.77-80.

Avnimelech, G., and Teubal, M. (2004). Venture Capital Start-Up co-evolution and the emergence and development of Israel's new high tech cluster. *Economic Innovation New Technologies*, 13 (1), pp. 33-60.

Ballarino, T., and Bellodi, L. (1997). *Gli aiuti di Stato nel diritto comunitario.* Napoli: Editoriale scientifica.

Balling, M., Bernet, B., and Gnan, E. (2009). *Financing SMEs in Europe,* Wien: SUERF.

Banca d'Italia, (2009). *Mezzogiorno e politiche regionali: Atti del Seminario.* Perugia: S.A.Di.Ba.

Barbieri, F., and Reggio, R. (2011). Sfuma il new deal delle Pmi. *Il Sole – 24 Ore*, 17 ottobre.

Bariatti, S. (1998). *Gli aiuti di Stato alle imprese nel diritto comunitario*. Milano: Giuffrè.

Bastagno, F. (1999). Il controllo comunitario degli aiuti di Stato nel recente regolamento di procedura. *Diritto del commercio internazionale*, 2, pp.339-372.

Benvenuto, L. (2005). La nozione di "microimpresa", di "piccola" e di "media impresa" negli aiuti alle attività produttive. *Corriere tributario*, 28, pp.2213-2218.

Biglieri, S. (1992). Aiuti alle piccole e medie imprese. *Diritto comunitario e degli scambi internazionali*, 4, pp.747-751.

Bilancia, F., and Filippini. A. (2011). L'intervento pubblico nell'economia delle Regioni e degli enti locali tra aiuti di Stati e tutela della concorrenza. *Le istituzioni del federalismo*, 3, pp.561-592.

Black, B.S., and Gilson, R.J. (1998). Venture Capital and the structure of capital markets: banks versus stock markets. *Journal of Financial Economics*, 47 (3), pp.243-277.

Brancati, R. (2010). *Fatti in cerca di idee: il sistema italiano delle imprese e le politiche tra desideri e realtà.* Roma: Donzelli.

Brancati, R. (2011). Stato e imprese: potenzialità e senso del ridicolo. www.nelmerito.com/.

Brancati, R., and Maresca, A. (eds.) (2010). *Le politiche per le imprese in Europa.* MET. Roma: Donzelli.

Bruno, E. (2009). Coesione europea e finanza per l'impresa. *Strategie di approccio e di gestione del sistema degli incentivi.* Milano: Giuffré.

Buzzacchi, C. (2004). Principio della concorrenza e aiuti di Stato tra diritto interno e diritto comunitario. *Giurisprudenza costituzionale*, 1, pp.277-291.

Calcagnini, G., and Favaretto, I. (eds) (2011). *L'economia della piccola impresa: Rapporto 2011.* Milano: F.Angeli.

Calabrese, G. (2001). La politica per la piccola e media impresa. In G. Vitali (a cura di), Imprese e mercati nell'Europa della moneta unica (pp. 245-273). Torino: UTET.

Cannari, L., D'Aurizio, L., and De Blasio, G. (2006). The effectiveness of investment subsidies: evidence from survey data. *Banca d'Italia: Occasional Papers*, n. 4.

Capelli, F., and Gratani, A. (1995). Aiuti e incentivi comunitari a favore delle piccole e medie imprese. *Diritto comunitario e degli scambi internazionali*, 3-4, pp.589-637.

Caprara, D., Carmignani, A., and D'Ignazio, A. (2010). Gli incentivi pubblici alle imprese: evidenza a livello micro. *Questioni di Economia e Finanza: Occasional Papers*, 60, pp.1-29.

Caputi Jambrenghi, V. (1998). Gli aiuti di Stato nel diritto comunitario vivente. *Rivista italiana di diritto pubblico comparato*, 6, pp.1259-1284.

Carabba, M. (1970). Incentivi finanziari, *Enc. Dir.*, XX. (pp. 963-973). Milano: Giuffrè.

Cardosi, E. (2010). Gli effetti delle agevolazioni finanziarie sulle decisioni d'investimento e sulle performance delle imprese: una valutazione empirica della legge 488/92. *Economia e diritto del terziario*, 1, pp.135-174.

Carullo, A. (2005). Lezioni di diritto pubblico dell'economia. Padova: Cedam.

Cassese, S. (2012). *La nuova Costituzione economica*. Roma:Laterza.

Castaldi, G. (1998). Sovvenzioni, incentivi e finanziamenti pubblici. Digesto: XV Commerciale (pp. 84-110). Torino: Utet.

Cavallo, B., and Di Plinio, G. (2007). Manuale di diritto pubblico dell'economia Milano: Giuffrè.

Center for Strategy and Evaluation Services, (2011). *Final evaluation of the entrepreneurship and innovation programme: Final report*. Sevenoaks: CSES.

Cocozza, F. (2007). *Diritto pubblico applicato all'economia.* Torino: Giappichelli.

Colin, F. (2011). Droit public économique. Sources et principes. Secteur public. *Régulation économique*. Paris: Gualino Editeur.

Commissione europea, (2009). Manuale delle norme comunitarie in materia di aiuti di stato a favore delle PMI. *Comprendente le misure temporanee di aiuto di stato a sostegno dell'accesso al finanziamento.* Bruxelles: 25 febbraio.

Commissione europea, (2008). Vademecum. Norme comunitarie applicabili agli aiuti di Stato. Bruxelles: 30 settembre.

Cossu, G. (1984). Finanziamenti: dir. pubbl.. *Enc. giur.*, XIV (pp. 1-6). Roma: Ist. Enc. It..

Cressy, R., and Olofsson, C. (1997). European SMEs financing: an overview. *Small Business Economics*, 9(2), pp. 87-96.

Croci, E. and Pericu, G. (1990). Sovvenzioni: diritto amministrativo. *Enc. del diritto*, XLIII (pp. 243-257). Milano: Giuffrè.

Czarnitzki, D. (2006). Research and development in small and medium - sized enterprises: the role of financial constraints and public funding. *Scottish Journal of political economy*, 53 (3), pp.335-357.

Damodaran, A. (2010). *Applied corporate finance*. New York: Wiley.

De Andreis, M., Maré, M. and Perissich, R. (2009). Le reazioni in tema di aiuti di Stato. In G. Amato (Ed.), *Governare l'economia globale* (pp. 335-351). Astrid, Firenze: Passigli Editori.

De Carli, P. (1995). *Lezioni e argomenti di diritto pubblico dell'economia*. Padova: Cedam.

Del Monte, A. (1994). Incentivi economici. *Enciclopedia delle Scienze sociali*, IV. (pp. 607-619). Roma: Treccani.

Di Comite, V. (2009). *Le sovvenzioni e le misure compensative nell'organizzazione mondiale del commercio*. Padova: Cedam.

Di Plinio, G. (1998). Diritto pubblico dell'economia. Milano: Giuffrè.

Dubocage, E., and Rivaud-Danset, D. (2002). Government policy on Venture Capital support in France. *Venture Capital*, 4 (1), pp.25-43.

Ecorys, (2011). Are EU SMEs recovering from the crisis? *Annual report on EU small and medium sized enterprises 2010/2011*. Rotterdam: European Commission.

European Central Bank, (2011). *Survey on the access to finance of small and medium – sized enterprises in the Euro area*: April to September 2011. Frankfurt: ECB.

European Commission, (2011). *Review of the Small Business Act for Europe*. Bruxelles.

European Commission - DG Enteprise and Industry, (2008). Communication from the Commission to the Council, the European Parliament, the European economic and social Committee and the Committee of the regions. *Think small first, a small business act for Europe*. Luxembourg: Office for official publications of the European Communities.

European Commission - DG Enteprise and Industry, (2010). European SMEs under Pressure. *Annual Report on EU Small and Medium-Sized Enterprises 2009.* Brussels.

European Commission - DG Research and Innovation-SME Unit, (2011). *Seventh Progress Report on SMEs participation in the 7th R&D Framework Programme*. Brussels.

Ferraro, F. (2010). L'evoluzione della politica sugli aiuti di Stato a sostegno dell'accesso al finanziamento nell'attuale situazione di crisi economica e finanziaria *Diritto dell'Unione europea*, 2, pp. 335-355.

Franchini, C. (1995). Piccola impresa e pubblici poteri. *Foro amm.*, 11-12, pp. 2823-2892.

Ganslandt, C. (2009). European Union SME policy. EU Finance Days for Smes. http://ec.europa.eu/small-business/index_it.htm.

Giannini, M. S. (1977a). Diritto pubblico dell'economia. Bologna: il Mulino.

Giannini, M. S. (1977b). Il finanziamento pubblico alle imprese con le risorse della collettività. *Giur. comm.*, 1, pp. 5-31.

Giannini, M. S. (1986). *Il pubblico potere: stati e amministrazioni pubbliche.* Bologna: Il Mulino.

Giusti, M. (1997). Lo Stato finanziatore: politiche ausiliative delle attività produttive. In M. Giusti (Ed.), Diritto pubblico dell'economia (pp.393-414), Padova: Cedam.

Goolsbee, A. (1998). Investment tax incentives, prices, and the supply of capital goods. *Quarterly Journal of Economics*, 113 (1), pp.121-148.

Hall, R.E., and Dale, W.J. (1967). Tax policy and investment behaviour. *American Economic Review*, 57 (3), pp.391-414.

Harding, R. (2002). Plugging the knowledge gap: an international comparison of the role for policy in the venture capital market. *Venture Capital*, 4 (1), pp.59-76.

Harris, R., and Trainor, M. (2005). Capital subsidies and their impact on total factor productivity: firm level evidence from northern Ireland. *Journal of Regional Science*, 45 (1), pp.49-74.

Huber, K. (1963). *Massnahmegesetz und Rechtsgesetz.* Berlin: Duncker and Humblot.

Iamiceli, P. (Ed.) (2009). *Le reti di imprese e i contratti di rete.* Torino: Giappichelli.

Jensen, M.C., and Meckling, W.H. (1976). Theory of the firm: managerial behavior, agency costs and ownership structure. *Journal of Financial Economics*, 3(4), pp.305-360.

Kaplan, S.N., and Stromberg, P. (2001). *Venture Capitalists as principals: contracting, screening and monitoring.* Cambridge: NBER WP 8202.

Kortum, S. and Lerner, J. (1998). *Does Venture Capital spur innovation?.* Cambridge: NBER WP 6846.

Lariccia, M. (2003). Sull'incompatibilità degli aiuti di Stato con il diritto comunitario. *Giustizia civile*, 2-3, pp.271-275.

Lawton, T.C. (2002). Missing the target: assessing the role of government in bridging the European equity gap and enhancing economic growth. *Venture Capital*, 4 (1), pp.7-23.

Lerner, J. (1999). The Government as Venture Capitalist: The long run impact of the SBIR program. *Journal of Business*, 72 (3), pp.285-318.

Loiodice, A. (1968). Revoca di incentivi economici ed eccesso di potere legislativo. In AA.VV., *Scritti degli allievi offerti ad Alfonso Tesauro* (pp.787-866). Milano: Giuffrè.

Lopriore, M. (2009). Supporting Enterprise Development and SME in Europe. *EIPASCOPE*, 2, pp.31-36.

Luchena, G. (2006). Aiuti pubblici e vincoli comunitari. Bari: Caccucci.

Malinconico, C. (1992). Tutela della concorrenza ed aiuti di stato nell'ordinamento comunitario. *Rivista trimestrale di diritto pubblico comunitario*, 2, pp. 431-469.

Malinconico, C. (2007). Aiuti di Stato. In M.P. Chiti and G. Greco (Eds.), *Trattato di diritto amministrativo europeo: parte speciale* (pp.66-115), I. Milano: Giuffrè.

Manzella, G. P. (2003). Gli ausili finanziari. In S. CASSESE (Ed.), *Trattato di diritto amministrativo: Diritto amministrativo speciale* (pp. 2861-2906), Milano: Giuffrè.

Mason, C.M., and Harrison, R.T. (1997). Business angel networks and the development of the informal venture capital market in the U.K.: is there still a role for the public sector?. *Small Business Economics*, 9(2), pp.111-123.

Ministero dello Sviluppo economico - Dipartimento per l'Impresa e l'Internazionalizzazione - Direzione Generale per le piccole e medie e gli enti cooperativi, (2011). Small Business

Act. *Le iniziative a sostegno delle PMI in Italia e nell'Europa a 27. Rapporto 2010.* Roma.

Ministero dello Sviluppo economico - Italia Istituto Nazionale Commercio Estero, (s.d.). *I programmi dell'UE a sostegno delle PMI e l'azione dell'ICE.* Roma.

Minniti, M. (2008). The role of Government policy on entrepreneurial activity: productive, unproductive or destructive?. *Entrepreneurship Theory and Practice*, 32 (5), pp.779–790.

Mortati, C. (1968). *Le leggi provvedimento.* Milano: Giuffrè.

Moscatelli, V. (2011). La cooperazione tra imprese e il contratto di rete. *Economia e diritto del terziario*, 1, pp.33-53.

Nannariello, G. (2012). I conti sugli incentivi alle imprese. www.lavoce.info.

OECD, (1997). *Government Venture Capital for technology-based firms.* Paris: OECD Publishing.

OECD, (2006). *The SME Financing Gap: Theory and Evidence.* Paris: OECD Publishing.

OECD, (2010). High-Growth Enterprises: What Governments Can Do to Make a Difference: OECD Studies on SMEs and Entrepreneurship. Paris: OECD Publishing.

OECD - Centre for Entrepeneurship, SMEs and Local Development, (2010). *The Impact of the Global Crisis on SME and Entrepeneurship Financing and Policy Responses.* Paris: OECD Publishing.

Orlandi, M. (1995). *Gli aiuti di Stato nel diritto comunitario.* Napoli: Edizioni Scientifiche Italiane.

Pace, A. (1983). Leggi di incentivazione e vincoli sul futuro legislativo. *Giur. Cost.*, I, pp.2341-2358.

Pace, A. (2004). Gli aiuti di Stato sono forme di "tutela" dalla concorrenza?. *Giurisprudenza costituzionale*, 1, pp.259-265.

Palmieri, C. (2002). In tema di sussistenza dei requisiti dimensionali per l'erogazione dei contributi a piccole e medie imprese. *Il Foro amministrativo: Tar,* 2, pp. 416-417.

Pappalardo, A. (1990). Aiuti regionali e controlli comunitari. Istituzioni comunitarie e regioni: effetti della normativa comunitaria sulle competenze regionali. Atti del 3° Convegno nazionale di studi regionali, Consiglio regionale della Liguria tenuto a Genova il 18-19 novembre 1988 (p. 113 – 118). Napoli: Edizioni scientifiche italiane.

Pedilarco, E. (2005). Piccole e medie imprese (PMI): il giudice comunitario chiarisce alcuni dubbi interpretativi. *Diritto pubblico comparato ed europeo*, I, pp.383-386.

Pericu, G. (1967). *Le sovvenzioni come strumento di azione amministrativa*: *I.* Milano: Giuffrè.

Picozza, E. (2001). *Il diritto pubblico dell'economia nell'integrazione europea.* Roma: Carrocci.

Pinna, G. (2004). La politica della concorrenza: Gli aiuti di Stato. *Rassegna di diritto pubblico europeo*, I, pp.145-170.

Pinotti, C. (2000). *Gli aiuti di Stato alle imprese nel diritto comunitario.* Padova: Cedam.

Porchia, O. (2004). La giurisprudenza comunitaria in materia di aiuti pubblici alle imprese. *Diritto pubblico comparato ed europeo*, 3, pp.1387-1395.

Pugliese, F. P. (1976). Problemi attuali della piccola media impresa. *Foro amm.*, 4, pp. 1124-1132.

Rabitti Bedogni, C. (2010). L'impatto della crisi economica sull'applicazione della normativa in materia di concorrenza e di aiuti di stato. *Relazione al Convegno "Crisi economico-finanziaria e disciplina degli aiuti di Stato".* Firenze: 20 settembre.

Roberti, G.M.(1992). Aiuti di Stato e controlli comunitari. *Foro italiano*, 10, 469-482.

Roberti, G.M. (1997). *Gli aiuti di Stato nel diritto comunitario*. Padova: Cedam.

Roccatagliata, F., and Medici, S. (1998). Normativa comunitaria in materia di aiuti concessi dallo Stato. *Rivista italiana di diritto pubblico comunitario*, 3-4, pp.619-643.

Sabatini, G. (2011). *Riesame dello Small Business Act: Audizione del Direttore generale dell'ABI presso la Camera dei Deputati*. Roma.

Sahlman, W.A. (1990). The structure and governance of Venture Capital organisations. *Journal of Financial Economics*, 27 (2), pp.473-521.

Santa Maria, A. (Ed.) (2006). *Concorrenza e aiuti di Stato*. Torino: Giappichelli.

Scaglione, M., and Caruso, B. (2001). Nuove tendenze nella disciplina degli aiuti di Stato nelle norme comunitarie. *Rivista italiana di diritto pubblico comunitario*, 2, pp.383-406.

Schepisi, C. (2011). La "modernizzazione" della disciplina sugli aiuti di Stato secondo l'Action Plan della Commissione europea: un primo bilancio. In C. Schepisi (Ed.), La "modernizzazione" della disciplina sugli aiuti di Stato: Il nuovo approccio della Commissione europea e i recenti sviluppi in materia di *public* e *private enforcement* (pp.17-34). Torino: Giappichelli.

Sediari, T., and Dell'Aquila, L.A. (2007). *Governance europea e strategie di sviluppo delle imprese*. Roma: Aracne.

Serrani, D. (1971). Lo stato finanziatore. Milano: F. Angeli.

Tassani, T. (2011). Il regime fiscale del 'nuovo' contratto di rete. In G. Calcagnini, and I. Favaretto (Eds.), *L'economia della piccola impresa: Rapporto 2011* (pp. 226-239). Milano: Franco Angeli.

Tonetti, A. (2003). Il finanziamento pubblico alle imprese: aspetti problematici consegnati al nuovo Titolo V della Costituzione. *Rivista giuridica del Mezzogiorno*, 2-3, pp.597-628.

Tortorelli, F. (1992). *Intervento pubblico e attività economiche*. Padova: Cedam.

Tosato, G. L. (2011). L'evoluzione della disciplina sugli aiuti di Stato. In C. Schepisi (Ed.), La "modernizzazione" della disciplina sugli aiuti di Stato: Il nuovo approccio della Commissione europea e i recenti sviluppi in materia di *public* e *private enforcement* (pp. 3-16). Torino: Giappichelli.

Trovato, P. G. (1993). Lo sviluppo della legislazione regionale di promozione economica e le situazioni soggettive degli operatori. In AA.VV., Il cittadino e la pubblica amministrazione dopo l'entrata in vigore del d.p.r. 24 luglio 1977 n. 616: *Atti del XXXVI Convegno di Studi di Scienza dell'amministrazione* (pp.191-225). Milano: Giuffrè.

Tufano, M. L. (2011). La disciplina degli aiuti di Stato dal controllo all'enforcement. In C. Schepisi (Ed.), La "modernizzazione" della disciplina sugli aiuti di Stato: Il nuovo approccio della Commissione europea e i recenti sviluppi in materia di *public* e *private enforcement* (pp.35-56). Torino: Giappichelli.

Text reviewed by Dr. Giovanni Dini (Manager of the Centro Studi Sistema of Ancona). Legal parts reviewed by Prof. Francesco Cocozza of the Department of Legal Sciences of the University of Ferrara).

In: Economics of Innovation, Incentives and Uncertainty ISBN: 978-1-62257-251-9
Editors: P. Simmons and S. T. Jordan

Chapter 2

REFORM ON BOARD OF DIRECTORS FOR TECHNOLOGY INNOVATION AND ECONOMIC SUCCESS IN JAPANESE MANUFACTURING FIRMS

***Toshiyuki Sueyoshi*[1,*], *Mika Goto*[2,†] and *Yusuke Omi*[3,‡]**

[1]New Mexico Institute of Mining and Technology,
Department of Management, Socorro, NM, US
[2]Central Research Institute of Electric Power Industry
Otemachi, Chiyodaku, Tokyo, Japan
[3]Tohoku University of Community Service and Science
School of Community Service and Science
Department of Community Service and Science,
Sakata, Yamagata, Japan

ABSTRACT

This study examines governance relationship between a board structure and corporate performance in Japan. It is easily imagined that a support from the top management is essential for both technical innovation and economic success in corporate activities. In other words, the board structure may determine technical innovation and economic success as the most influential management component. In this research, the firms sampled for this study consist of 280 Japanese manufacturing firms. The first section of Tokyo Stock Exchange Market lists all the firms examined in this study. The observed period is from 2003 to 2006. The empirical results partly support the effectiveness of the ongoing efforts on a board structure reform in Japanese firms. This study identifies the three empirical findings. First, a size reduction in the board does not increase the corporate performance of Japanese manufacturing firms. Second, the introduction of independent directors enhances the corporate performance. Finally, the high independency of directors further improves the corporate performance.

---

* E-mail: toshi@nmt.edu.
† E-mail: mika@criepi.denken.or.jp.
‡ E-mail: omi@koeki-u.ac.jp.

Keywords: Corporate Governance, Board of Directors, Independent Directors, Technology Innovation

INTRODUCTION

A major role of the top management in firms is that it initiates and supports possible organizational and technology innovation within their firms, thereby leading to their economic successes. In such a major role of the top management, a board of directors needs to monitor their corporate performance. The board provides corporate leaders with various types of advices from their monitoring results. Such a governance effort can enhance the corporate performance of firms. Therefore, a firm needs to organize strategically an effective board of directors to enhance the corporate performance.

Recently, many foreign investors, in particular institutional investors, request Japanese firms to reform the boards of directors because they believe that the board reform is necessary for Japanese firms to enhance the corporate performance of their firms. To respond to their requests, Japanese firms attempt to reduce the number of board members and to use an independent director(s) as their board reform strategies (Sueyoshi et al, 2010).

Acknowledging the importance of Japanese board reform, this study has been long wondering whether the board reform really improves the corporate performance of Japanese manufacturing firms because foreign investors, not by themselves, mainly suggest the board reform. To examine the Japanese governance issue in manufacturing firms, this study is interested in investigating the following three research issues: First, does a size reduction in a board of directors enhance the corporate performance of Japanese manufacturing firms (Yermack, 1996, Bhagat and Black 1999)? Second, does the introduction of an independent director enhance their corporate performance (Mehran, 1995, Agrawal and Knoeber, 1996)? Finally, besides the two current governance issues in Japanese manufacturing firms, this study is interested in how to deal with an "endogeneity problem", often existing in research on corporate governance. As discussed by many previous studies (e.g., Hermalin and Weisbach, 1988, Denis and Sarin, 1999, Linck et al., 2008 and Guest, 2008, Jackling and Johl, 2009), the structure for a board of directors (i.e., the size of board members and the number of independent directors) influences the corporate performance. It is also true that the corporate performance influences the structure of a board of directors. The causality issue may produce an opposite result on our first and second research concerns. Research on corporate governance has long discussed an existence of the endogeneity problem. However, no previous study has sufficiently documented how to handle the problem, in particular from the perspective of corporate governance issues in Japanese manufacturing firms.

Considering the three research agendas discussed above, this study is concerned with empirical investigation concerning how the reform on a board of directors really enhances the corporate performance, measured by Tobin's q ratio, of Japanese manufacturing firms. This study investigates Japanese corporate governance, which is different from those of American and European firms (e.g., Pindado and Torre, 2011). Not many studies have explored current governance issues in Japanese manufacturing firms. The firms selected for this study have more than 100 billion (yen: Japanese currency) in terms of their total sales and total assets.

All of them are listed in the first section of the Tokyo Stock Exchange Market. The number of firms is 280. The sampled period is from 2003 to 2006.

The remainder of this study is organized as follows: The next section reviews previous research efforts on a board of directors and corporate performance. This literature review specifies the position of this study by comparing it with the other previous studies. This section also summarizes hypotheses examined in this study. The proceeding sections describe research strategy used in this study and summarize our statistical and empirical results. The last section concludes this study along with future research agendas.

I. Literature Review and Hypotheses

A. Conventional View on Japanese Corporate Governance

The research of Kang and Shivdasani (1995) discussed Japanese corporate governance, firm performance and top executive turnover based upon the five characteristics of Japanese firms. The five characteristics, distinguishing Japanese firms from American and European ones, included (a) keiretsu (an industrial group), (b) main bank system, (c) ownership structure, (d) board of directors and (e) corporate acquisitions. It is known that many previous studies have discussed Japanese corporate governance based upon the five characteristics (so, five grouping). Therefore, this study fully utilizes the five characteristics proposed by Kang and Shivdasani (1995). The initial task of this study is to update the literature review by adding recent studies on Japanese corporate governance. After classifying the recent literature into the five groups, this study describes the position of this research by comparing it with the updated previous studies.

Keiretsu (an industrial group): An important feature of traditional Japanese corporate governance was that many large firms belonged to an industrial group (*keiretsu*). A major commercial bank served as a central organization in the group and played a leading role of financial activities within the group. Firms belonging to a *keiretsu* were bound together by a nexus of explicit and implicit contacts and maintained substance business ties with other firms in the group (e.g., Kang and Shivdasani, 1999, and Mueller, 2006). Thus, the previous studies considered that Japanese firms were affiliated to *keiretsu* whose affiliation status was more hybrid in nature and more independent than American and European firms. The aspect of *keiretsu* was historically true when we discussed Japanese manufactring firms.

Main bank: Japanese firms traditionally depended upon main bank financing. Many Japanese corporations had a commecial bank, often referred to as "a main bank", among their major stockholders. The main bank provided both long-term and short-term financing and played a major role in obtaining operational capital for its client firms. The main bank was responsible on monitoring firm performance and often intervened when profitablity was insufficent. The previous studies considered that the main bank system, often within *keiretsu*, characterized the Japanese corporate governance (e.g., Miwa and Ramseyer, 2005 and Suetorsak, 2007).

Ownership structure: Japanese financial institutions held a large fraction of voting rights in many firms. According to the anti-monopoly act, Japanese banks could invest in up to 5% of the equity of a firm. Japanese banks and insurance firms owned 42.2% of the shares listed on Tokyo Stock Exchange in 1987. Such concentrated ownership could provide Japanese financial institutions with incentives to monitor corporate performance of firms within their groups. Consequently, Japanese financial institutions were influential on corporate investment decisions (e.g., Anderson and Campbell, 2004 and Mueller, 2006).

Board of directors: Outside directors were generally believed to perform an important corporate governance function in the United States. In contrast, a majority of Japanese corporate directors were long-term employees and they were board members for a substantial length of time. The observed number of outside directors was relatively low, but not completely absent. It was suggested by many previous studies that outside directors were often appointed to improve insufficient corporate performance (e.g., Miwa and Ramseyer, 2005; and Mueller, 2006).

Corporate acquisitions: Many studies discussed that corporate takeovers played an important role in disciplining American managers whose performance was insufficient. In contrast, corporate acquisitions were a less important governance role in Japan (Miwa and Ramseyer, 2005 and Schneper and Guillen, 2004).

The review of previous studies provides us with the historical fact on Japanese corporate governance and firm performance. However, these previous studies did not docunment the current corporate governance in Japan. Such new aspects can be found in the following two major changes in Japanese governance:

One of the two changes is that after the collapse of the bubble economy, the *keireitsu* does not fuction as it used to be. For example, the center of *keiretsu* was a large commerical bank. Mitsubishi bank mergered with Tokyo bank and they became Tokyo-Mitsubishi bank. Sanwa bank mergered with Tokai bank and they became UFJ bank. Now, the two bank groups mergered to become Mitsubishi-Tokyo-UFJ bank. In the merger process of Japanese banking system, the *keiretsu* has become gradually less important in Japan because firms in a *keiretsu* group find other competing firms in a same industry within the same group. Thus, firms cannot find a major business benefit to stay in their *keiretsu* group. The change in *keiretsu*'s business role also indicates that a main bank system does not function as before. Furtheremore, some major banks went bankrupt (e.g., Hokkaido Takushoku Bank, Long-Term Credit Bank of Japan and Nippon Credit Bank). The remaining other banks were reluctant to extend new lending and withdrew their lending. Thus, it was very difficult for Japanese firms to maintain the main bank system, as discussed in the previous studies. These business changes in the last decade imply that both the *keiretsu* and the main bank system, as summarized above, do not explain the current relationship between firm performance and corporate governance in Japan.

The other important aspect on current Japanese governance can be found in the statistics of Tokyo Stock Exchange (TSE). According to the survey of share ownership (http://www.tse.or.jp/market/data/examination/distribution/ index.html), foreign shareholdings rose from 3.9% in 1989 to 22.2% in 2005. Another TSE statistics (Trading Volume by Investor Type TSE 1st section (http: www.tse.or.jp/english/market/data/sector/ index.html) indicated that foreign investors accounted for approximately 55% of total consignment trade regarding the number of shares traded in 2007. Paying attention to an individual firm, the percentage of shares held by foreign investors of TDK rose from 12.24% (1990) to 40.97% (2006) and Sony from 18.78% (1990) to 48.09% (2006). Thus, it is not appropriate to examine the relationship between corporate governance and firm performance without considering the influence of foreign investors. Unfortunately, the previous studies have insufficiently examined the important aspect on Japanese corporate governance. Finally, note that Bae et al. (2006) have confirmed the important of this concern on foreign investors in TSE market.

B. New Research Task on Japanese Corporate Governance

Among the five characteristics of Japanese corporate governance, this study pays attention to the board of directors, hereafter, because foreign investors are most interested in the board reform of Japanese firms to improve a return from their investments. To examine the governance issue, this study focuses upon the following two issues: (a) a size on board of directors and (b) a level of independency of independent directors. The two governance issues are essential components of the board reform in Japan.

A Size on Board of Directors: A unique feature of Japanese corporate governance is that it often consists of a larger board of directors than that of American and European firms. This business aspect is because many employees often stay long in Japanese firms and a director position is an incentive for them as the final goal of their internal promotion system. Consequently, Japanese firms need to prepare many positions within a board of directors in order to acknowledge employees' long contributions.

The large board of directors yields a positive result in Japanese firms, but it may simultaneously produce a negative result. For example, Lipton and Lorsch (1992) and Jensen (1993) described that a large board of directors disrupted communication among board members and their speedy decision-making. Yermack (1996) and Eisenberg et al. (1998) provided empirical evidence that the reduction of a board size enhanced corporate performance. See Bhagat and Black (1999, 2001) and Mak and Li (2001). In a similar research effort, Coles et al. (2008) examined the relationship between a board size and corporate performance (measured by Tobin's q ratio) by using 8165 samples obtained from 1992 to 2001. They found a negative relationship between them. Thus, the previous studies have indicated that the reduction in a board of directors does not immediately imply the enhancement in corporate performance of all firms. The inconclusive suggestions indicate a research opportunity that we need to investigate the current governance trend in Japanese firms because they try to reduce the size of board members by the suggestion of foreign investors.

Next, paying attention to Japanese corporate governance, almost no research has explored on a linkage between the size of a board of directors and the corporate performance within Japanese firms. An exception was Miwa and Ramseyer (2005). Their study examined the relationship between the number of board members and corporate performance which was measured by various financial indexes such as Tobin's q ratio and ROE (Return on Equity) in Japanese firms. Their finding indicated that an influence of a board size depended upon its performance measure. Thus, their research did not conclude on the influence of a board size.

In addition to the governance issues, it is important to mention that many previous studies have insufficiently explored an endogeneity problem in their empirical investigations. As discussed by Denis and Sarin (1999), the endogeneity problem existed in research on corporate governance in such a manner that the corporate performance of firms determined the size of a board of directors. In particular, firms with low profitability reduced the board size and increased the number of independent directors.

Based upon the literature review summarized above, this study is interested in the following first hypothesis:

- *Hypothesis 1:* The reduction in a board of directors increases the corporate performance of Japanese manufacturing firms under the treatment of an endogeneity problem.

Independent Director: An expected role of an independent director includes monitoring current corporate performance of a firm, preventing a self-centered corporate use by CEO and other managers, and protecting a benefit of equity holders. Unfortunately, the expected role may not function in many Japanese firms because board members usually consist of employees who have been working in a firm for a long period. In many cases, a CEO has power to designate and dismiss directors of a board. Thus, it is difficult for board members to express a negative opinion to the CEO. They always make consensus building among members so that they easily exclude a negative opinion on current management and performance. Consequently, the board of directors may not function for monitoring and advising within their firm.

To overcome the governance problem, foreign investors suggest that Japanese firms should use an independent director who can monitor corporate performance, being independent from the other board members. Here, it is important to specify an independent director as an individual who was not an employee of a firm. He can advise for a firm from a strategic perspective that is different from other executives, as discussed by Boot et al. (2005). However, the independent director has a difficulty in accessing information enough to utilize his independent view as a board member. An increase in the number of independent directors may decrease a firm's strategic decision-making capability. Thus, it is important for a firm to balance between independent and other board members (Tirole, 2005).

From the above concerns, we classify the previous studies into two groups of research results. One of the two groups included Yermack (1996), Luan and Tang (2007) and Chan and Li (2008) who indicated a positive influence of an independent director on corporate performance measured by Tobin's q ratio and ROE. The other group included Agrawal and Knoeber (1996) and Bhagat and Black (1999, 2001), who indicated an opposite result. Thus, the influence of an independent director on corporate performance is still inconclusive until now.

Returning to the endogeneity problem, this study confirms that the problem exists in this type of research group (e.g., Hermalin and Weisbach, 1988 and Denis and Sarin, 1999). For example, Linck et al. (2008) reported that inadequate performance in 6931 US firms invited an increase in the number of independent directors.

The literature review discussed above provides the following hypothesis to be examined in this study.

- *Hypothesis 2:* The introduction of an independent director(s) increases the corporate performance of Japanese manufacturing firms under the treatment of an endogeneity problem.

II. RESEARCH STRATEGY

A. Model Structure and Variable Selection

The endogeneity problem is specified by the following simultaneous equations:

$$Performance = f\ (Lboard, Indep, Z)\ , \quad (1)$$

$$Lboard = f\ (Performance, Z) \text{ and} \quad (2)$$

$$Indep = f\ (Performance, Z). \quad (3)$$

Here, *Lboard* indicates the number of board members in a natural logarithm. *Indep* indicates the ratio of independent directors to all board members. Z is a set of exogenous variables that influence the corporate performance of firms and their board structures.

An important feature of Equations (1)-(3) is that the board structure (*Lboard* and *Indep*) of a firm influences the corporate performance, as specified in (1). In contrast, the corporate performance influences the board structure (e.g., the number of board members and the number of independent directors), as expressed by (2) and (3), along with exogenous variables.

In addition to the endogeneity problem, this study needs to mention the following five concerns on Equations (1) to (3):

First, this study uses Simple q ratio as a measure of corporate performance in Equations (1)-(3). The Simple q ratio corresponds to the ratio of the market value of a firm to the replacement cost of its physical assets at the end of each year. To calculate the ratio, this study divides the sum of the market value of shareholders' equity and the book value of debt by the book value of total assets.

Second, the exogenous variables (Z) include *Size*, *Dar*, *RandD ratio* and *Adv ratio*. *Size* indicates a corporate size because this study expects that the size of a firm has a positive effect on its performance. Furthermore, a large firm tends to have a large board of directors (Boone et al, 2007). *Dar* indicates the ratio of total debt to total assets. As discussed by Coles et al. (2008) and Linck et al. (2008), when a firm faces a difficulty in managing its financial stability because of a high ratio of debt, the firm often increases the number of financial directors who can provide professional advices on the debt reduction. *RandD ratio* indicates a ratio of RandD expenditure to total assets. An increase in the ratio may enhance corporate performance by new product developments and new patents (Guest, 2008). *Adv ratio* is the ratio of advertizing cost to total assets. It is a straight expectation that an increase in the ratio enhances corporate performance by establishing the brand image of a firm. However, as discussed by Linck et al. (2008), detailed information concerning RandD expenditure and advertizing cost usually exists within a firm so that an independent director cannot access the important information. Consequently, the independent director cannot provide the firm with appropriate monitoring and advices, so reducing the size of *Indep*.

Third, this study incorporates *Dirown*, indicating a stock option, as an exogenous variable of Z because the availability of a stock option system may positively influence the corporate performance of a firm. As discussed by Jensen and Meckling (1976), when managers, board members and shareholders can share a same goal, they may attain an alignment effect among them so that the stock option may produce a positive influence on corporate performance. In addition, the stock option may increase a bargaining power of board members so that it enhances corporate performance (Mak and Li, 2001). Acknowledging the importance of such a positive effect, this study needs to mention another aspect on the stock option. That is, if a manager and/or a director own a large share of equity, he may control a firm under his rule and management style. Such dictatorship may produce a negative effect on his firm's corporate performance (McConnell and Servaes, 1995).

Fourth, this study uses efficient two-step GMM (Generalized Method of Moments) as the estimation methodology in order to identify relationship between corporate performance and *Lboard*/*Indep* under a possible occurrence of the endogeneity problem. Dummy variables are used to indicate different annual periods and different industries in the proposed GMM estimation. The GMM is a general statistical method that is widely used in econometrics. See Cameron and Trivedi (2005) that provided a detailed description on GMM.

Finally, the proposed two-step GMM uses instrumental variables, besides the exogenous variables, that have a linkage with the structure of a board of directors, but not having any direct linkage with corporate performance. Following Barnhart and Rosenstein (1998) and other previous studies, this study employs one-period and two-period lag variables for endogenous and exogenous variables as instrumental variables.

In addition to these variables mentioned above, this study uses one- and two-period lag variables of *CEOtenure* and *Retvar* as instrumental variables. The *CEOtenure* indicates a tenure period on how long a president stays in a firm. The longer the president stays in his firm, the more influential power he has on the board of directors and he tends to avoid introducing an independent director (Hermalin and Weisbach, 1998). Meanwhile, no previous studies use *CEOtenure* to explain his firm performance, concluding that there is no clear evidence to indicate a direct linkage between *CEOtenure* and corporate performance. The *Retvar* indicates the variance of a return from equity. This study incorporates it as the instrumental variable because it is difficult for a firm to monitor its corporate performance when the equity of a firm exhibits high volatility (so, high variance) in a stock market. In the case, the firm needs to make a special group of executives within a board of directors in order to monitor the corporate performance. Thus, the variance of stock return influences the board structure of directors. Meanwhile, no previous study considered *Retvar* as an explanatory variable for measuring corporate performance. Thus, there is no clear evidence to indicate a direct relationship between *Retvar* and coporate performance.

B. Data Description

This study sampled a data set on board members and independent directors from "Yakuin-Shikihou" (implying "Quarterly Report on Executives" in Japanese) of Touyou Keizai Publisher. We also sampled a data set on the other variables from "Kigyou Zaimu Data Bank" (implying "Corporate Finance Data Bank in Japanese) from Japanese Economic Research Center.

In sampling the data sets, this study pays attention to only Japanese manufacturing firms, listed in Tokyo Stock Exchange Market, which have more than 100 billion (Japanese Yen) in terms of the total amount of assets and the total amount of sales. Furthermore, we examine the data availability in such a manner that we can sample a data set on these firms during at least 10 consecutive years. Consequently, this study obtains 280 firms in a balanced panel data structure. The total number of samples is 2800 (= 10 x 280). The data sets used for this study are summarized in Table 1 as descriptive statistics, including average, S.D. (Standard Deviation), Min. (Minimum), Max. (Maximum), Median, 25% and 75% percentiles and IQ (Inter-Quartile) range.

Table 1. Descriptive Statistics

	Average	S.D.	Min.	Max.	Median	25 Percentile	75 Percentile	IQ Range
Simple q	1.55	0.69	0.52	10.00	1.35	1.13	1.73	0.60
Lboard	2.43	0.42	1.10	3.61	2.40	2.08	2.71	0.63
Indep	0.07	0.12	0	0.79	0	0	0.11	0.11
Number of Independent Directors	0.76	1.32	0	11	0	0	1	1
Number of Directors	12.40	5.57	3	37	11	8	15	7
Size	19.64	0.92	17.90	23.09	19.46	18.95	20.14	1.19
Dar	0.49	0.18	0.00	0.94	0.49	0.37	0.63	0.26
R & D ratio	0.04	0.03	0.00	0.24	0.02	0.01	0.05	0.04
Adv ratio	0.01	0.02	0.00	0.12	0.00	0.00	0.01	0.01
Dirown	0.91	3.14	0.00	35.78	0.11	0.05	0.31	0.25
Ceotenure	1.31	0.86	0.00	3.97	1.39	0.69	1.79	1.10
Retvar	0.01	0.01	0.00	0.26	0.00	0.00	0.01	0.01

Simple q: (Total Market Value = Book Value of Liability with Interest) / Book Value of Total Asset.
Lboard: Natural Logarithm of Number of Directors.
Indep: Number of Independent Directors / Number of Total Directors.
Size: Natural Logarithm of Total asset.
Dar: Total Liability / Total Asset.
Dirown: Share of Number of Stocks Held by Directors (%).
R & D ratio: R and D Expenditures / Total Asset.
Adv ratio: Advertisment Expenditures / Total Asset.
CEOtenure: Natural logarithm of Length of Service.
Retvar: Variance of Monthly Rate of Return on Equity Investment During Last 12 Months.

Before describing empirical results, this study needs to mention the current trend on Japanese corporate governance. For example, Table 2 summarizes an annual shift in the number of board members from 2003 to 2006. The table indicates that the number of board members has reduced from 13.2 on average in 2003 to 11.8 in 2006. Thus, there was a declining trend in the number of board members in the observed period. Furthermore, the ratio of independent directors increased from 6.2% on average in 2003 to 8% in 2006. Thus, the ratio exhibited an increasing trend. The two trends suggest that the reform on a board of directors in Japanese manufacturing firms reduces the size of board members and increases an independent director(s) within the board. However, the introduction of an independent director is still limited in Japanese manufacturing firms because the number of firms, having the independent director, was 118 in the sampled 280 firms (2006).

Table 2. Time Shift in Structure of Board Members

Year	Number of Total Directors	Share of Independent Directors (%)	Number of Firms with Independent Directors	Numbr of Firms
2003	13.2	6.2	99	280
2004	12.7	6.7	103	280
2005	12.2	7.3	105	280
2006	11.8	8.0	118	280

III. STATISTICAL TESTS

Figure 1 visually describes the relationship among the three equations (1)-(3), including exogenous and instrumental variables, along with statistical tests and their purposes.

Endogeneity: The arrow above (1) indicates a causal effect expressed by Equation (1). This study is interested in examining whether the causal effect exists or not in the corporate governance of Japanese manufacturing industry. A methodological issue associated with the causality is that there is the other causal effect depicted by the arrow below (2) and (3) in Figure 1 and expressed by Equations (2) and (3) as discussed by many previous studies on corporate governance. This study uses the Durbin-Wu-Hausman test to examine whether the causal effect expressed by (2) and (3) exists or not in this study. If the causal effect does not exist in this study, it is possible for us to apply OLS (Ordinary Least Squares) to estimate linear regression between corporate performance (as a dependent variable) and *Lboard/Indep* (explanatory variables). In an opposite case where we need to consider the causal effect expressed by (2) and (3), the OLS usually produces biased and inconsistent estimates. Thus, it is impossible to use the OLS in this study. However, if appropriate instrumental variables are available, we can obtain consistent estimates to express a linear regression in the coexistence of (1), (2) and (3).

Identification Test on Instrumental Variables: An instrumental variable is causally associated with a board structure, which consists of endogenous variables (*Lboard* and *Indep*) to be instrumented, but the variable is not directly associated with corporate performance. There is still a case that the instrumental variable has high correlation with corporate performance where a single source of such a correlation is an indirect path of the instrumental variable via the board structure. The indirect pass determines corporate performance. To identify whether instrumental variables are valid and necessary, this study uses two identification tests: (a) Kleibergen-Paap rank Wald Test to examine their weak identification and (b) Kleibergen-Paap rank LM Test to examine their underidentification. Here, the underidentification implies that instrumental variables are irrelevant and the model may be unidentified if a limited number of instruments are relevant. Meanwhile, the weak identification implies that instrumental variables can identify the board structure, but they are insufficient in terms of their identification capability. This is the case when a very low correlation exists between the instrument and the endogenous variable being instrumented (*Lboard* and *Indep*). In the case, the model is weakly identified and the instrument is called a weak instrument. The identification tests correspond to the arrow from "Instrumental Variables" to "Board Structure" in Figure 1.

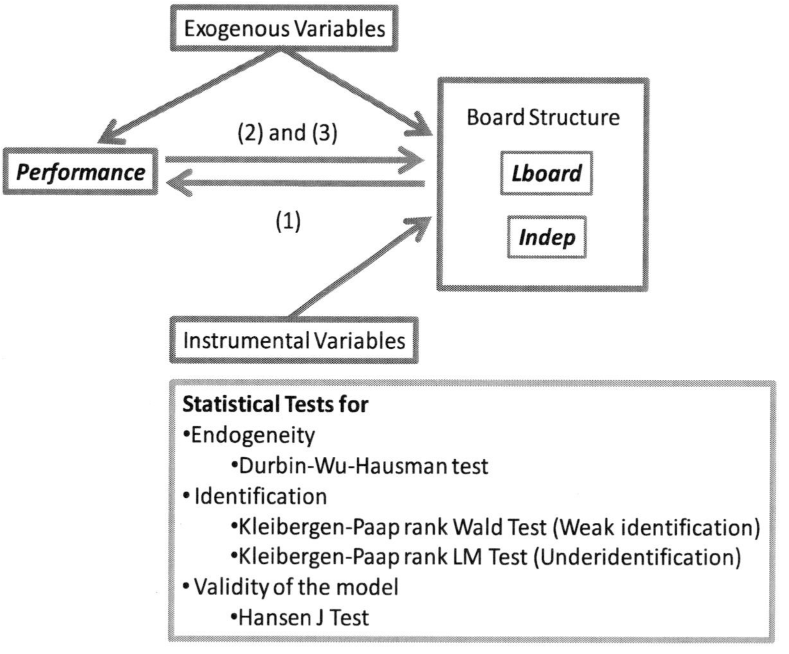

Figure 1. Structure of Estimation Model Used in This Empirical Study. Note: (a) Null hypothesis (Test for Endogeneity) is whether an OLS estimator of the same equation yields consistent estimates. Instrumental variables techniques are not required. (b) Null hypothesis (Test for Identification) is whether the rank condition for identification is not satisfied. (c) Null hypothesis (Test for Validity of the Model) is whether moment conditions of a model are satisfied.

Validation of Whole Model: This study needs to examine whether the proposed (whole) model, visually summarized in Figure 1, centers in the causal relationship expressed by the arrow (1) under the treatment of the endogeneity problem with exogenous and instrumental variables. We use Hansen J test to examine the assumption regarding whether the proposed model fits well or not the data set used in this study. The Hansen J test, or J test, is often referred to as a test for overidentification.

IV. ESTIMATION RESULTS

Table 3 summarizes estimation results of six regression models measured by the two-step GMM. The differences among these regression models are due to a combination among instrumental and exogenous variables. This study examines such different combinations to confirm the robustness of the estimation results. The importance of such a robustness check was suggested by Barnhart and Rosenstein (1998).

To make a statistical linkage between Figure 1 and Table 3, Table 4 summarizes resulting statistical tests on six different models. As listed in the top of Table 4, this study has applied the Durbin-Wu-Hausman test to examine whether *Lboard* and *Indep* are endogenous variables. All models listed in Table 4 indicate that the two variables are endogenous at the level of 1% significance so that estimated coefficients are different between OLS and the two-step GMM. Note that OLS results are not listed in Table 3 because the estimation is biased. Therefore, the difference indicates that this study needs to use the two-step GMM method to examine a causal effect from the two variables (*Lboard* and *Indep*) to corporate performance, as depicted by the arrow (1) of Figure 1, under the treatment of endogeneity problem.

In addition to the Durbin-Wu-Hausman test, this study examines the problem of identification and the validity of models used in this study, as listed in Figure 1. First, this study examines whether the identification problem exists on all the models, using the Kleibergen-Paap rank Wald F-ratio (weak identification) and Kleibergen-Paap rank LM statistic (underidentification). See the middle rows on Table 4.

Table 3. Estimation Results by Two-Step GMM

	Dependent Variable : *Simple q*					
	Model 1	Model 2	Model 3	Model 4	Model 5	Model 6
Explanatory Variables	Coef. Est.	Coef. Est.	Coef. Est.	Coef. Est.	Coef. Est.	Coef. Est.
Lboard	0.061	0.099	0.202 **	0.041	0.080	0.181 *
	(0.09)	(0.11)	(0.10)	(0.09)	(0.10)	(0.10)
Indep	1.036 *	1.768 ***	2.086 ***	0.724	1.352 **	1.676 ***
	(0.57)	(0.67)	(0.48)	(0.49)	(0.62)	(0.45)
Size	0.056	0.024	-0.001	0.070 **	0.035	-0.001
	(0.04)	(0.04)	(0.04)	(0.04)	(0.04)	(0.04)
Dar	-1.356 ***	-1.206 ***	-1.363 ***	-1.352 ***	-1.186 ***	-1.342 ***
	(0.17)	(0.20)	(0.20)	(0.17)	(0.20)	(0.19)
R and D Ratio		1.619	1.193		1.785	1.455
		(1.19)	(1.12)		(1.14)	(1.10)
Adv ratio		1.676	3.093 **		1.629	3.331 **
		(1.50)	(1.38)		(1.48)	(1.37)
Dirown		0.030 **	0.020 *		0.026 **	0.016
		(0.01)	(0.01)		(0.01)	(0.01)
Constant	0.636	1.020	1.280 *	0.430	0.854	1.310 *
	(0.64)	(0.72)	(0.69)	(0.61)	(0.70)	(0.67)
Instrument Variables						
Lboard (1 – and 2 – period lag)	√	√	√	√	√	√
Indep(1 – and 2 – period lag)	√	√	√	√	√	√
Size(1 – and 2 – period lag)			√			√
Dar(1 – and 2 – period lag)			√			√
R & D Ratio(1 – and 2 – period lag)			√			√
Adv ratio(1 – and 2 – period lag)			√			√
Dirown(1 – and 2 – period lag)			√			√
CEOTenure(1 – and 2 – period lag)				√	√	√
Retvar(1 – and 2 – period lag)				√	√	√

Simple q: (Total Market Value = Book Value of Liability with Interest) / Book Value of Total Asset.
Lboard: Natural Logarithm of Number of Directors.
Indep: Number of Independent Directors / Number of Total Directors.
Size: Natural Logarithm of Total asset.
Dar: Total Liability / Total Asset.
Dirown: Share of Number of Stocks Held by Directors (%).
R & D ratio: R and D Expenditures / Total Asset.
Adv ratio: Advertisment Expenditures / Total Asset.
CEOtenure: Natural logarithm of Length of Service.
Retvar: Variance of Monthly Rate of Return on Equity Investment During Last 12 Months.
(a) Estimated coefficients of year dummy and industry dummy variables are omitted from this table.
(b) Numbers in parentheses indicate a robust standard error. Superscripts ***, **, and * indicate significance at the level of 1%, 5%, and 10%, respectively.
(c) Symbol of ✔ indicates the instrumental variables used under the models.

Table 4. Statistical Tests of Models

Statistical Tests			Model 1	Model 2	Model 3	Model 4	Model 5	Moedl 6
Endogenity	Durbin-Wu-Hausman Test		88.65 **	193.89 **	84.23 **	94.1 ***	44.69 ***	47.56 ***
Identification	Kleibergen-Paap rank Wald F-ratio		188.32	226.85	73.68	99.61	120.75	62.66
	Stock and Yogo (2005): Critical Value	5 %	11.04	11.04	19.83	17.70	17.70	20.33
		10%	7.56	7.56	10.89	10.22	10.22	11.00
		20%	5.57	5.57	6.20	6.20	6.20	6.14
		30%	4.73	4.73	4.53	4.73	4.73	4.43
	Kleibergen-Paap rank LM statistics		182.96 ***	150.47 ***	152.43 ***	183.36 ***	151.32 ***	153.37 ***
Validity of the model	Hansen J Test		3.28 (0.19)	1.80 (0.41)	17.09 (0.15)	8.52 (0.20)	8.13 (0.23)	25.24 (0.06)
	Number of Samples		560	438	429	560	438	429

(a) The numbers in parentheses of Hansen J Test indicate the p-values.
(b) Cameron and Trivedi (2005) and Greene (2008) described all statistical tests listed in Table 4.

Considering the weak identification problem, this study finds that a bias in weak instruments, compared to that of OLS, is less than 5%. The bias is much smaller than 20%, or a general threshold proposed by Stock and Yogo (2005), a value used to test the problem of weak identification. Thus, the problem of weak identification is limited on all the models listed in Table 4. On the problem of underidentification, the null hypothesis is rejected at the level of 1% significance. Thus, no model examined in Table 4 has the problem of underidentification. The two tests imply that instrumental variables used in this study have an estimation capability to identify the board structure (i.e., *Lboard* and *Indep*).

Second, this study examines the validity of all models including instrumental and exogenous variables, using the Hansen J test that examines the moment conditions for GMM estimation. The null hypothesis of the Hansen J test is whether the moment conditions required for GMM estimation are satisfied. We expect that the hypothesis should not be rejected for the validity of the model. As summarized in Table 4, Model 6 is rejected at the level of 10% but the other models are not rejected. Thus, all the models with an exception of Model 6 are valid and useful for our empirical study.

Since this study has completed the statistical tests in Table 4, we return to Table 3 in order to interpret parameter estimates of the six models. Model 3 and 6 have a positive sign on *Lboard* at the 5% level of significance, but the other models do not show any significance on the coefficient estimate. Model 6 does not pass the validity test by the Hansen J test. Consequently, only Model 3 among the six models shows that the size of a board (*Lboard*) has a positive effect on corporate performance. Thus, the assertion does not have its robustness. In other words, the empirical result obtained in this study is inconsistent with the research of Yermarck (1996) which has confirmed that the size reduction in a board of directors has a positive effect on corporate performance. Meanwhile, the estimation results listed in Tables 3 and 4 indicate that the size reduction in a board does not enhance the corporate performance of Japanese manufacturing firms. Such a result is consistent with the research of Mak and Li (2001). Thus, the first hypothesis is invalid in this study.

Next, this study pays attention to the coefficient of *Indep* in Table 3. The coefficient of all models, except Model 4, is positive at least at the level of 10% significance. This result indicates that the introduction of an independent director enhances corporate performance. This result is different from previous studies such as Agrawal and Knoeber (1996), Bhagat and Black (2001) and Miwa and Ramseyer (2005). However, even if we consider a possible endogeneity problem, we can empirically confirm that the independent director is useful in enhancing corporate performance. Thus, the second hypothesis is valid.

To investigate the influence of an independent director further, this study estimates its elasticity, finding that the elasticity becomes 0.06 in Model 5. This result implies that an increase of 1% in *Indep* increases 0.06% (= 1.352 x 0.07/1.55) in Simple q ratio. To explain the result further, let us consider a firm whose board of directors consists of 10 members. If the firm increases the number of independent directors from 2 to 3, then the ratio of independent directors to all directors increases 50%. The increase enhances 3% in Simple q ratio under the condition that the amount of total assets is same before and after the appointment of an independent director. Thus, an independent director enhances corporate value by 3%, implying a considerably large improvement on corporate performance.

The preceding section has confirmed that an independent director has a positive effect on the enhancement in corporate performance. To investigate the governance issue further, this study pays attention to a degree of independency of an independent director. It is easily imagined that high independency is important for an independent director because he cannot provide an appropriate monitoring capability on corporate performance and its related advice capability without high independency.

Here, it is important for this study to define the "high independency" of an independent director. The "Director Independency Code" prepared by Japanese Association of Board of Directors (2010), specifies "high independency" by the following five requirements: (a) an independent director is not listed as one of ten large equity holders; (b) he has never worked for a bank with which his firm has currently received a financial service; (c) he has never worked for a firm which is one of ten large equity holders; (d) he has never worked for affiliate companies of his firm; and (e) he is not a director of a firm that has an interlocking directorate with his firm.

Based upon the definition on high independency, this study separates *Indep* into its related two ratio measures. One of the two measures is *HIndep* that indicates the ratio of independent directors with high independency who satisfy the above five requirements to all board members. The other measure is *LIndep* that indicates the ratio of independent directors with low independency to all board members.

The low independent director implies an individual who does not have high independency. Table 5 summarizes descriptive statistics on *HIndep* and the number of high independent directors. Table 6 shows its annual shift from 2003 to 2006. The two tables indicate an increasing trend in *HIndep*.

Table 5. Descriptive Statistics on High Independent Directors

	Average	S.D.	Min.	Max.	Median	25 percentile	75 Percentile	IQ Range
Numebre of Higher Independent Directors	0.47	0.98	0	8	0	0	1	1
HIndep	0.04	0.09	0	0.58	0	0	0.05	0.05

(a) HIndep: Number of Higher Independent Directors/Number of Total Directors.

Table 6. A Time Shift of High Independent Directors

Year	Numbe of higher Independent Directors	Share of Higher Independent Directors (%)	Number of Firms with Higher Independent Directors	Number of Toatal firms
2003	0.39	3.7	67	280
2004	0.44	4.2	73	280
2005	0.47	4.4	75	280
2006	0.58	5.4	90	280

Table 7. Influence of High Independent Director on Corporate Performance

	Dependent Variable : *Simple q*					
	Model 1	Model 2	Model 3	Model 4	Model 5	Model 6
Explanatory Variables	Coef. Est.	Coef. Est.	Coef. Est.	Coef. Est.	Coef. Est.	Coef. Est.
Lboard	0.070	0.097	0.178 *	0.045	0.071	0.145
	(0.09)	(0.10)	(0.10)	(0.9)	(0.10)	(0.10)
HIndep	1.246 *	1.966 **	2.292 ***	0.933	1.430 *	1.558 **
	(0.71)	(0.92)	(0.70)	(0.62)	(0.87)	(0.64)
Size	0.057	0.023	0.001	0.067 *	0.036	0.011
	(0.04)	(0.04)	(0.04)	(0.04)	(0.04)	(0.04)
Dar	-1.384 ***	-1.220 ***	-1.382 ***	-1.363 ***	-1.194 ***	-1.341 ***
	(0.17)	(0.20)	(0.20)	(0.17)	(0.20)	(0.20)
R & D Ratio		1.779	1.511		1.889	1.837
		(1.25)	(1.17)		(1.20)	(1.15)
Adv ratio		1.468	3.113 **		1.531	3.297 **
		(1.55)	(1.44)		(1.53)	(1.43)
Dirown		0.030 **	0.021 *		0.028 **	0.019 *
		(0.01)	(0.01)		(0.01)	(0.01)
Constant	0.662	1.117	1.389 *	0.509	0.903	1.238 *
	(0.67)	(0.75)	(0.72)	(0.62)	(0.74)	(0.70)
Instrument Variables						
Lboard (1 – and 2 – period lag)	√	√	√	√	√	√
HIndep(1 – and 2 – period lag)	√	√	√		√	√
Size(1 – and 2 – period lag)			√			√
Dar(1 – and 2 – period lag)			√			√
R & D Ratio(1 – and 2 – period lag)			√			√
Adv ratio(1 – and 2 – period lag)			√			√
Dirown(1 – and 2 – period lag)			√			√
CEOTenure(1 – and 2 – period lag)				√	√	√
Retvar(1 – and 2 – period lag)				√	√	√

Simple q: (Total Market Value = Book Value of Liability with Interest) / Book Value of Total Asset.
Lboard: Natural Logarithm of Number of Directors.
HIndep: Number of Higher Independent Directors / Number of Total Directors.
Size: Natural Logarithm of Total asset.
Dar: Total Liability / Total Asset.
Dirown: Share of Number of Stocks Held by Directors (%).
R & D ratio: R and D Expenditures / Total Asset.
Adv ratio: Advertisment Expenditures / Total Asset.
CEOtenure: Natural logarithm of Length of Service.
Retvar: Variance of Monthly Rate of Return on Equity Investment During Last 12 Months.
(a) Estimated coefficients of year dummy and industry dummy variables are omitted from this table.
(b) Numbers in parentheses indicate a robust standard error. Superscripts ***, **, and * indicate significance at the level of 1%, 5%, and 10%, respectively.
(c) Symbol of ✔ indicates the instrumental variables used under the models.

Table 7 summarizes parameter estimates of the six models after incorporating *HIndep* instead of *Indep*. Table 7 confirms that *HIndep* has a statistically significant positive effect on corporate performance in all models. An exception is Model 4.

Table 8 uses the Hansen J test to examine the validity of the six models under the case of Table 7. As listed in Table 8, the test rejects the null hypothesis on Model 3 at the level of 10% significance and that of Model 6 at the level of 5% significance for *HIndep*. All the other statistical tests of Table 8, including the Hansen J test, have indicated the models with *HIndep* are valid.

Table 9 summarizes parameter estimates of the six models after incorporating *LIndep* instead of *Indep*. Table 10 uses the Hansen J test to examine the validity of the six models under the case of Table 9. As listed in Table 10, the test rejects the null hypothesis on Model 3 at the level of 10% significance and that of Model 6 at the level of 5% significance for *LIndep*. All the other statistical tests in Table 10, including the Hansen J test, have indicated the models with *LIndep* are valid.

Table 8. Statistical Tests of Models

Statistical Tests		%	Mod. 1	Mod. 2	Mod. 3	Mod. 4	Mod. 5	Mod. 6
Endogenity	Durbin-Wu-Hausman Test		75.72 ***	34.06 *	40.97 **	74.14 ***	46.11 ***	97.08 ***
Identification	Kleibergen-Paap rank Wald F-ratio		185.13	244.94	76.82	107.32	130.08	66.22
	Stock and Yogo(2005): Critical Value	5	11.04	11.04	19.83	17.70	17.70	20.33
		10	7.56	7.56	10.89	10.22	10.22	11.00
		20	5.57	5.57	6.20	6.20	6.20	6.14
		30	4.73	4.73	4.53	4.73	4.73	4.43
	Kleibergen-Paap rank LM statistics		182.96 ***	150.88 ***	150.81 ***	185.74 ***	152.82 ***	152.78 ***
Validity of the model	Hansen J Test		4.49 (0.11)	3.86 (0.15)	20.32 (0.06)	9.32 (0.16)	9.43 (0.15)	28.01 (0.03)
	Number of Samples		560	438	429	560	438	429

(a) The numbers in parentheses of Hansen J Test indicate the p-values.

As listed in Table 9, *LIndep* has a statistically significant positive effect on corporate performance in all models. Exceptions are Models 1 and 4. Thus, the introduction of an independent director enhances the corporate performance, not depending upon whether they are *HIndep* or *LIndep*, as found in Tables 7 and 9.

Admitting the importance of the business implication found in Tables 7 and 9, this study needs to add that the elasticity of *HIndep* in Model 5 is 0.04, larger than 0.03 of *LIndep*. Thus, *HIndep* has more influence on corporate performance than *LIndep*. A similar result can be found in the other models. Thus, it is better for Japanese manufacturing firms to introduce an independent director, in particular with high independency.

Table 9. Influence of Low Independent Director on Corporate Performance

	Dependent Variable : *Simple q*					
	Model 1	Model 2	Model 3	Model 4	Model 5	Model 6
Explanatory Variables	Coef. Est.	Coef. Est.	Coef. Est.	Coef. Est.	Coef. Est.	Coef. Est.
Lboard	-0.007	-0.041	0.111	0.004	0.008	0.118
	(0.09)	(0.12)	(0.10)	(0.08)	(0.10)	(0.09)
LIndep	0.675	2.013 ***	1.955 ***	0.205	1.521 **	1.633 ***
	(0.82)	(0.77)	(0.61)	(0.70)	(0.70)	(0.59)
Size	0.083 ***	0.074 **	0.049	0.090 ***	0.072 **	0.040
	(0.03)	(0.04)	(0.03)	(0.03)	(0.04)	(0.03)
Dar	-1.355 ***	-1.043 ***	-1.269 ***	-1.362 ***	-1.067 ***	-1.280 ***
	(0.17)	(0.21)	(0.20)	(0.17)	(0.21)	(0.20)
R & D Ratio		3.072 ***	2.893 **		2.957 ***	3.160 ***
		(1.15)	(1.13)		(1.14)	(1.12)
Adv ratio		1.003	2.316 *		1.082	2.745 **
		(1.44)	(1.37)		(1.41)	(1.34)
Dirown		0.037 ***	0.036 ***		0.034 ***	0.029 ***
		(0.01)	(0.01)		(0.01)	(0.01)
Constant	0.352	0.353	0.569	0.201	0.293	0.708
	(0.56)	(0.63)	(0.62)	(0.54)	(0.63)	(0.61)
Instrument Variables						
Lboard (1 – and 2 – period lag)	√	√	√	√	√	√
LIndep(1 – and 2 – period lag)	√	√	√	√	√	√
Size(1 – and 2 – period lag)			√			√
Dar(1 – and 2 – period lag)			√			√
R and D Ratio(1 – and 2 – period lag)			√			√
Adv ratio(1 – and 2 – period lag)			√			√
Dirown(1 – and 2 – period lag)			√			√
CEOTenure(1 – and 2 – period lag)				√	√	√
Retvar(1 – and 2 – period lag)				√	√	√

Simple q: (Total Market Value = Book Value of Liability with Interest) / Book Value of Total Asset.
Lboard: Natural Logarithm of Number of Directors.
LIndep: Number of Lower Independent Directors / Number of Total Directors.
Size: Natural Logarithm of Total asset.
Dar: Total Liability / Total Asset.
Dirown: Share of Number of Stocks Held by Directors (%).
R & D ratio: R and D Expenditures / Total Asset.
Adv ratio: Advertisment Expenditures / Total Asset.
CEOtenure: Natural logarithm of Length of Service.
Retvar: Variance of Monthly Rate of Return on Equity Investment During Last 12 Months.

Table 10. Statistical Tests of Models

Statistical Tests			Model 1	Model 2	Model 3	Model 4	Model 5	Moedl 6
Endogenity	Durbin-Wu-Hausman Test		76.94 ***	59.09 ***	75.52 ***	138.61 ***	98.56 ***	119.8 ***
Identification	Kleibergen-Paap rank Wald F-ratio		39.91	103.96	32.34	21.40	63.11	29.53
	Stock and Yogo (2005): Critical Value	5 %	11.04	11.04	19.83	17.7	17.7	20.33
		10%	7.56	7.56	10.89	10.22	10.22	11
		20%	5.57	5.57	6.2	6.2	6.2	6.14
		30%	4.73	4.73	4.53	4.73	4.73	4.43
	Kleibergen-Paap rank LM statistics		36.46 ***	33.16 ***	42.21 ***	40.14 ***	39.81 ***	45.44 ***
Validity of the model	Hansen J Test		3.40 (0.18)	0.96 (0.62)	19.17 (0.08)	7.76 (0.26)	6.49 (0.37)	26.22 (0.05)
	Number of Samples		560	438	429	560	438	429

(a) The numbers in parentheses of Hansen J Test indicate the p-values.

Conclusion and Future Extensions

Following governance strategy suggested by foreign investors, Japanese firms recently reduced the size in a board of directors and introduced an independent director(s) to attain speedy decision-making, efficient monitoring and effective advice. Such a new business trend is important for innovation and economic success of Japanese firms. To examine the validity of the recent governance strategy, this study examined corporate governance (measured by a board structure) and firm performance (measured by Tobin's q ratio), considering a possible occurrence of an endogeneity problem. This study used the two-step GMM as an estimation methodology under the treatment of an endogeneity problem. Previous studies on corporate governance have not sufficiently explored the treatment of an endogeneity problem. On the best of our knowledge, this study is the first research effort to overcome the methodological difficulty from the perspective of an effective board reform of directors in major Japanese manufacturing firms. This study applied the proposed approach to 280 Japanese manufacturing firms, all of which were listed in the first section of Tokyo Stock Exchange Market. The observed period was from 2003 to 2006, which was a transitional period for the board reform in corporate governance of Japanese firms.

This study identified the following three important empirical findings: First, this study could not statistically confirm that the size reduction in a board of directors enhanced the corporate performance of Japanese manufacturing firms. Second, the introduction of independent directors increased their corporate performance. Finally, high independence of independent directors enhanced the corporate performance of firms. The first finding does not support the suggestion from foreign investors, but the second and third findings support the validity of their suggestions.

This study has the following three drawbacks, all of which we need to overcome in future. First, we have sampled only large manufacturing firms in Japan with high corporate values. Therefore, the three findings in this study may not be applicable to the smaller manufacturing firms and the other industries. The business implications obtained in this study is thus limited as scientific evidence. To overcome the drawback, it is necessary for us to extend the scope of this study to other firms whose sizes are less than the firms examined in this research. Such an extension also includes non-manufacturing industries such as service industry (Venkatraman, 1994). Second, this study does not pay attention to an optimal size of a board of directors and that of independent directors. The research extension is another future task of this study. Finally, this study does not investigate governance issues related to race, age and gender diversifications because old and male executives dominate a board of directors in Japanese manufacturing industry. The Japanese firms consist of an old boy network. However, as discussed by Kang et al. (2007), Japanese firms cannot escape from the diversification issue. Japanese firms must compete in a global market where the diversification is very important as long as they want to operate in a global market. For example, Japanese auto-industry must pay attention to the race and gender diversification in the United States. Otherwise, they violate the U.S. law on EEO (Equal Employment Opportunity). Thus, it is important for this study to extend itself toward the diversification issue as an important future research agenda.

Finally, it is hoped that this study contributes for research on corporate governance. We look forward to seeing future extensions as discussed in this study.

References

Agrawl, A., Knoeber, C.R. 1996. Firm performance and mechanisms to control agency problems between managers and shareholders. *Journal of Financial and Quantative Analysis,* 31: 377-397.

Anderson, C.W., Campbell, T.L. 2004. Corporate governance of Japanese banks, *Journal of Corporate Finance*, 10: 327-354.

Bae, K.-H., Yamada, T., Ito, K. 2006. How do individual, institutional and foreign investors win and lose in equity trades?: Evidence from Japan. International Review of Finance, 6: 129-155.

Barnhart, S.W., Rosenstein, S. 1998. Board composition, managerial ownership, and firm performance, an empirical analysis. *The Financial Review*, 33:1-16.

Bhagat, S., Black, B. 1999. The uncertain relationship between board composition and firm performance. *Business Lawyer*, 54: 921-963.

Bhagat, S., Black, B. 2001. The non-correlation between board independence and long term firm performance. *Journal of Corporation Law*, 27: 231-274.

Boone, A.L., Field, L.C., Karpoff, J.M., Raheja,C.G. 2007. The determinants of corporate board size and composition: an empirical analysis. *Journal of Financial Economics*, 85: 66-101.

Boot, W.A., Macey, J.R., Schmeits, A.B., 2005. Towards a Nnw theory of corporate governance: objectivity versus proximity, Available at SSRN: http://ssrn.com/abstract =686770.

Cameron, A.C., Trivedi, P.K. 2005. *Microeconometrics: Methods and Applications*, Cambridge University Press, New York.

Chan, K.C., Li, J. 2008. Audit committee and firm value: Evidence on outside top executives as expert-independent directors. *Corporate Governance: An International Review*, 16: 16-31.

Coles, J., Daniel, N., Naveen, L. 2008. Boards: does one size fit all? *Journal of Financial Economics,* 87: 329-356.

Denis, D.J., Sarin, A. 1999. Ownership and board structures in publicly traded corporations. *Journal of Financial Economics*, 52: 187-223.

Eisenberg, T., Sundgren, S., Wells, M.T. 1998. Larger board size and decreasing firm value in small firms. *Journal of Financial Economics*, 48: 35-54.

Guest, P.M. 2008. The determinants of board size and composition: Evidence from the UK. *Journal of Corporate Finance*, 14: 51-72.

Hermalin, B. E., Weisbach, M.S. 1988. The determinants of board composition. *The RAND Journal of Economics*, 19: 589-606.

Hermalin, B.E., Weisbach, M.S. 2003. Boards of directors as an endogenously determined institution: A survey of the economic literature. *Federal Reserve Bank of New York Economic Policy Review*, 7:7-26.

Jackling, B., Johl, S. 2009. Board structure and firm performance: Evidence from Idia's top campanies. *Corporate Governance: An International Review*, 17: 492-509.

Jensen, M.C. 1993. The modern industrial revolution, exit, and the failure of internal control systems. *Journal of Finance*, 48: 831-880.

Jensen, M.C., Meckling, W.H. 1976. Theory of the firm: managerial behavior, agency costs, and ownership structure. *Journal of Financial Economics*, 3: 305-360.

Kang, H., Cheng, M., Gray, S.J. 2007. Corporate governance and board composition: Diversification and independence of Australian boards. *Corporate Governance: An International Review*, 15: 194-207.

Kang, J.K., Shivdasani, A. 1995. Firm performance, corporate governance and top executive turnover in Japan, *Journal of Financial Economics*, 38, 29-58.

Luan, C.J., Tang, M.J. 2007. Where is independent director efficacy?" *Corporate Governance: An International Review*, 15: 636-643.

Linck, J.S., Netter, J.M., Yang,T. 2008. The determinants of board structure. *Journal of Financial Economics*, 87: 308–328.

Lipton, M., Lorsch, J.W. 1992. A modest proposal for improved corporate governance. *Business Lawyer*, 48: 59-77.

Mak, Y., Li, Y. 2001. Determinants of corporate ownership and board structure: Evidence from Singapore. *Journal of Corporate Finance*, 7: 231-256.

McConnell, J.J., Servaes, H. 1995. Equity ownership and the two faces of debt. *Journal of Financial Economics*, 39:131-157.

Mehran, H. 1995. Executive compensation structure, ownership, and firm performance. *Journal of Financial Economics*, 38: 163-184.

Miwa,Y., Ramseyer, J.M. 2005. Who appoints them, what do they do? Evidence on outside directors from Japan. *Journal of Economics and Management Strategy*, 14: 299–337.

Mueller, D.C. 2006. Corporate governance and economic performance, *International Review of Applied Economics*, 20: 623-643.

Pinado, J., Torre, C. 2011. Capital Structure: New evidence from the ownership structure. *International Review of Finance*, 11: 213-226.

Schneper, W.D., Guillen, M.F. 2004. Shareholder rights and corporate governance: a cross-national study of hostile takeovers, *Administrative Science Quarterly*, 49: 263-295.

Sueyoshi, T., Goto, M., Omi, Y. 2010. Corporate governance and firm performance: Evidence from Japanese manufacturing industries after the lost decade. *European Journal of Operational Research*, 203: 724-736.

Suetorsak, R., 2007. Keiretsu and risk: an examination of the risk exposure of keiretsu banks in Japan, *Journal of Economics and Finance*, 31: 268-282.

Tirole, J., 2005. *The Theory of Corporate Finance*, Princeton University Press.

Venkatraman, N., Loh, L., Koh, J. 1994. The adoption of corporate governance mechanisms: A test of competing diffusion models. *Management Science*, 40: 496-507.

Yermack, D. 1996. Higher market valuation of companies with a small board of director. *Journal of Financial Economics*, 40: 185-211.

In: Economics of Innovation, Incentives and Uncertainty ISBN: 978-1-62257-251-9
Editors: P. Simmons and S. T. Jordan

Chapter 3

A MODELLING APPROACH FOR THE EVALUATION OF NONPOINT POLLUTION POLICY IN THE PRESENCE OF UNCERTAINTY

Graeme J. Doole[1,2] and Upananda H. Paragahawewa[3]
[1]Centre for Environmental Economics and Policy, School of Agricultural and Resource Economics, Faculty of Natural and Agricultural Sciences,
University of Western Australia, Crawley, Western Australia, Australia
[2]Department of Economics, Waikato Management School, University of Waikato,
Private Hamilton, New Zealand
[3]AgResearch, Ruakura Research Centre, Hamilton, New Zealand

ABSTRACT

Water quality throughout the world is declining in response to diffuse pollution from agricultural production. It is difficult to highlight cost-effective means to assess alternative policy instruments for its regulation because of the presence of much uncertainty regarding annual pollutant flows, agent heterogeneity, climate and soil characteristics, and future price movements. Bioeconomic modelling represents a flexible means to assess the suitability of different policy instruments in the presence of such uncertainty. A general framework that considers alternative on-farm mitigation technologies in different land uses in agricultural catchments is introduced for the analysis of such problems. The model is applied to the Lake Taupo catchment in New Zealand to identify the abatement costs associated with emissions standards of different intensities and the relative value of market-based and uniform policy instruments. Emission reductions beneath 30 per cent are found to incur minimal abatement costs when appropriate mitigation technologies are adopted. This reflects the low abatement cost present in the simulated dairy sector and the cost-effective mitigation practices available to achieve reasonable leaching reductions on sheep and beef farms, which constitute more than 90 per cent of the catchment. Moreover, this finding is robust to broad changes in key model parameters. Thus, land use heterogeneity is highlighted as an important determinant of whether such policies are appropriate for the regulation of agricultural activity to improve water quality.

Keywords: Abatement cost, market-based policy, nitrogen, nonpoint source pollution

1. Introduction

The identification of cost-effective means to regulate nonpoint source (NPS) pollution is difficult because of the presence of much uncertainty regarding annual pollutant flows, agent heterogeneity, climate and soil characteristics, and future price movements. Nonetheless, the cost of different policies is important to determine, as agriculture plays an important role in many economies, especially those of developing countries. Much work in the area of nonpoint pollution regulation currently focuses on the use of cooperative agreements among producers to secure reductions in pollution based on observed ambient emissions (Segerson, 1988; Suter et al., 2008). These involve rewards/penalties being levied for farmers in a catchment when ambient water quality is above/below a given threshold. These instruments directly address the fact that ambient water quality is influenced by stochastic impacts that are not directly attributable to producers and NPS is a collective issue arising from a population from which individual responsibility is difficult to ascribe. Nonetheless, these instruments are only suitable where there are a small number of farmers and changes in ambient water quality are directly attributable to farmer actions, rather than being dominated by stochastic processes (Weersink et al., 1998).

Accordingly, this paper takes an alternative focus. It addresses uncertainty through the construction of a bioeconomic model that may be calibrated to represent a broad range of agricultural catchments across the world. The uncertain parameters and relationships that characterise the NPS pollution problem are thus represented quantitatively within a decision making framework in which the implications of alternative policy instruments can be assessed. Moreover, the difficulties associated with stochastic modelling (Doole and Kingwell, 2010) are overcome, with a focus placed on deterministic optimisation and structured sensitivity analysis. The method is demonstrated in the context of a case study involving Lake Taupo in New Zealand.

Previous authors have used bioeconomic models to assess NPS pollution policies (e.g. Helfand and House, 1995). However, the modeling framework presented here is important because:

1. It overcomes the difficulties associated with the identification of appropriate stochastic processes to include in decision models of NPS pollution (Doole and Kingwell, 2010).
2. It allows the use of a decision making framework to test the impacts of alternative assumptions regarding the magnitude of different processes in the model. For example, the impacts of changes in the profitability of sheep and beef farming on abatement cost in the case study application are shown below.
3. There is no general economic framework for NPS pollution policy evaluation in a watershed incorporating multiple agricultural land uses.
4. The model has a clear economic interpretation.

5. Its level of complexity can be manipulated depending on the quality and quantity of available data.
6. It can be easily optimised using any nonlinear programming (NLP) solver (Bazaraa et al., 2006; Brooke et al., 2008).
7. The structure of the model allows the efficient identification of global optima, without recourse to stochastic-search algorithms.
8. It provides a convenient description of alternative abatement technologies by avoiding the computational difficulty associated with the representation of multiple binary control variables when many discrete mitigation alternatives exist.

Lake Taupo is one of New Zealand's iconic natural assets. It is of high recreational value, possessing a mature tourist industry with annual revenue of around $NZ400 million. Lake Taupo also provides hydroelectric power and is a source of drinking water for Taupo township, which contains around 21,000 people. However, there are strong indications from water quality modelling and ongoing monitoring (Elliott and Stroud, 2001; Gibbs, 2008) that increasing nitrate levels pose an impending threat to the provision of the primary values associated with Lake Taupo. Current regulatory policy has been effective in limiting point source emissions. Thus, consistent with global trends, the primary threat is nitrate leaching from agricultural production. This is evident in the background level of nitrogen in Lake Taupo, which is currently 3–5 times greater than natural levels due to agricultural development (Hadfield, 2007). The nitrogen excreted by grazing livestock is the primary source of these emissions, with 80–90 per cent of nitrogen ingested by grazing cattle being expelled in dung and urine (Monaghan et al., 2007).

Expansion of dairy farming into the Lake Taupo catchment has been a critical threat to future water quality given the high prices received for milk production in New Zealand in recent years (Vant and Huser 2000) and the high levels of nitrate (generally above 35 kg N ha^{-1}) typically leached in these systems (Monaghan et al., 2007). This stimulated the introduction of a cap-and-trade policy for nitrate emissions in this catchment in late 2007. The overall goal of the policy is to achieve a 20 per cent emissions reduction, a target estimated to achieve long-term stabilisation of water quality in Lake Taupo at a reasonable level. All farms in the catchment area are allocated emission allowances based on their highest level of emissions from 2004–07. Regulators expect allowance trading among NPS polluters, but none has occurred thus far. The Lake Taupo Protection Trust has been charged with achieving the 20 per cent emissions reduction through the funding of research, buying emissions allowances, and purchasing farms and converting them to less-intensive land uses, such as production forestry.

Market-based instruments have been widely applied in developed economies given their theoretical promise to mitigate the market failure that underlies negative environmental externalities (Newell and Stavins, 2003; De Cara et al., 2005). Nonetheless, uniform regulations are valuable alternatives, as they can avoid problems associated with high transactions costs, initial entitlement allocation, and thin permit markets (Helfand and House, 1995; Newell and Stavins, 2003). This study estimates the abatement costs associated with the regulation of agricultural production within the Lake Taupo catchment in a large number of policy settings. This is important since there is an apparent dearth of economic analysis that predicts the costs associated with different emissions targets, potential costs of the current policy, and the relative value of alternative approaches, particularly uniform regulations.

The paper is structured as follows. Section 2 introduces a general framework for policy evaluation in agricultural catchments incorporating diverse land uses. Section 3 describes the application of this model to the Lake Taupo catchment. Section 4 presents results of policy evaluation and sensitivity analysis. Section 5 concludes.

2. THEORETICAL MODEL

This section presents a model incorporating alternative on-farm mitigation technologies in different land uses within a catchment.

Insights gained are important for the Lake Taupo case specifically, but also provide valuable general results given the pervasive impact of agriculture on water quality throughout Australasia (Drewry et al., 2006) and the world (United Nations Environment Program, 2008).

The model is described as follows. Assume a catchment consists of four types of pastoral land-use. This number may be easily altered, but four are relevant for the purposes of this application. These sectors are: bull beef (denoted *B*), dairy cows (denoted *C*), deer (denoted *D*), and sheep and beef (denoted *S*) production.[1] The total land area allocated to each of the four activities is described through the intervals $i = [1,2,..., I]$, $j = [1,2,..., J]$, $k = [1,2,..., K]$, and $l = [1,2,..., L]$ for bull beef, dairy, deer, and sheep and beef farming, respectively. Here, *I*, *J*, *K*, and *L* denote the total number of hectares allocated to each sector and $I + J + K + L = T$, where *T* is the total land area of the catchment. The upper bounds $\{I, J, K, L\}$ are set at current levels. However, each farming activity can switch to a land-use that emits fewer nitrates as a mitigation option. This sharpens the focus on efficient regulation of the intensive margin, in accordance with the standard NPS pollution literature (Shortle and Horan, 2001).

The level of intensification within the bull beef, dairy, deer, and sheep and beef sectors is defined through the decision variables N_i^B, N_j^C, N_k^D, and N_l^S, respectively. These denote the per-hectare level of nitrate leaching selected on hectare *i*, *j*, *k*, and *l* in the corresponding sector as the relevant emissions choice given a particular set of economic, regulatory, and technical circumstances. The level of production intensity chosen in each land use is associated with net returns through a concave profit function. The functions used in each sector are defined $f_i^B(N_i^B)$, $f_j^C(N_j^C)$, $f_k^D(N_k^D)$, and $f_l^S(N_l^S)$ for bull beef, dairy, deer, and sheep farming, respectively, where $f'(\cdot) > 0$ and $f''(\cdot) < 0$. The curvature of these profit functions is consistent with the stylised facts of production theory and previous modelling of the relationship between stocking intensity and profit on pastoral farms (e.g. Doole, 2008).

A logistic profit function is used in this application given its consistency with these curvature assumptions and powerful description of reality. Four such functions are defined, one for each land use. Their common form is:

---

[1] Interdependency between sectors is ignored due to a lack of information. It also has minimal impact on the key conclusions due to its limited scale and since it would only serve to further decrease abatement costs and the degree of heterogeneity.

$$f(N) = gN\left(1 - \frac{N}{X}\right), \tag{1}$$

where g is the slope of the profit function, N is the level of farm intensification, and X is the maximum level of potential emissions in each land use. Profit increases with intensification on the left-hand side of the curve (Figure 1), as typically observed in New Zealand agriculture (Monaghan et al., 2007). In contrast, production declines beyond some optimal threshold because of nutrient toxicity and constraints associated with fixed capital. For example, higher stocking rates can increase nitrate leaching, but simultaneously decrease profit because of soil compaction and overgrazing. These mechanisms will eventually force farm profit to zero, denoted here as $N = X$ (Figure 1).

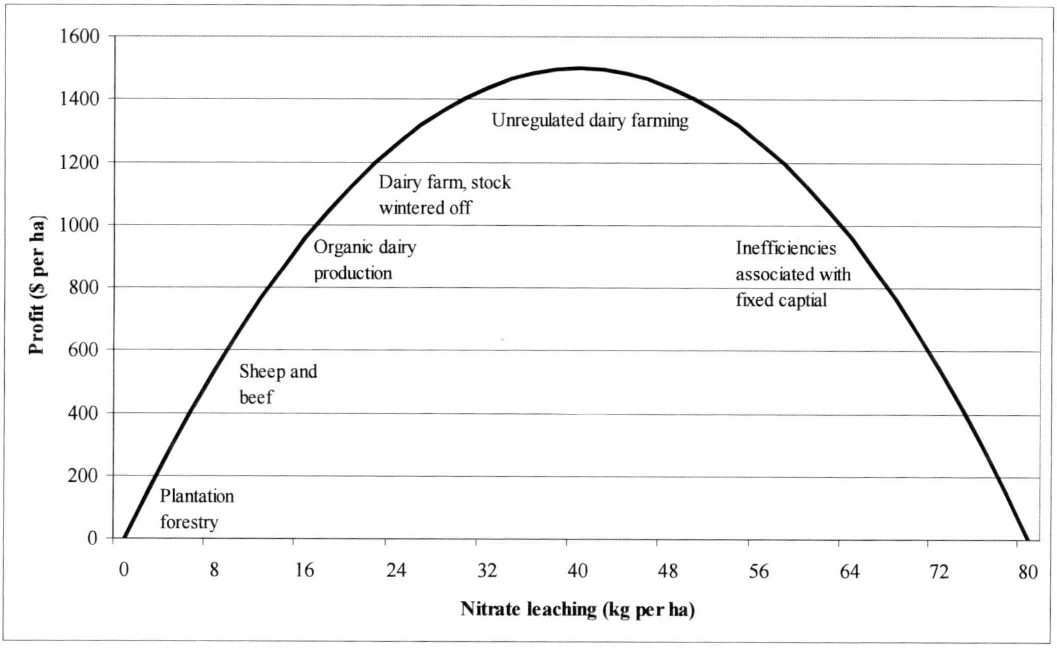

Figure 1. Hypothetical logistic profit function for dairy production representing relationship between mitigation options and farm returns.

Individual profit functions can be defined per-hectare, per-farm, or per-sector. Information requirements increase as this relationship is defined at finer scales. Thus, the per-sector approach is used here due to the lack of information. This is likely to be most typical given a general dearth of information regarding the relationship between nitrate leaching and profit specified per-hectare or per-farm (Shortle and Horan, 2001).

In an unregulated environment, the maximum profit derived from the catchment (Z) is the solution to the unconstrained problem:

$$\max_{N_i^B, N_j^C, N_k^D, N_l^S} Z = \sum_{i=1}^{I} g^B N_i^B \left(1 - \frac{N_i^B}{X_i^B}\right) + \sum_{j=1}^{J} g^C N_j^C \left(1 - \frac{N_j^C}{X_j^C}\right) + .$$

$$+\sum_{k=1}^{K} g^D N_k^D \left(1 - \frac{N_k^D}{X_k^D}\right) + \sum_{l=1}^{L} g^S N_l^S \left(1 - \frac{N_l^S}{X_l^S}\right) \tag{2}$$

The optimal level of intensity for each hectare within each land-use can be determined through standard calculus (Bazaraa et al., 2006), which yields a set of first-order conditions (FOCs):[2]

$$\frac{\partial Z}{\partial N_i^B} = g^B \left(1 - \frac{2N_i^B}{X_i^B}\right) = 0 \rightarrow N_i^B = \left(X_i^B / 2\right) \ \forall i, \tag{3}$$

$$\frac{\partial Z}{\partial N_j^C} = g^C \left(1 - \frac{2N_j^C}{X_j^C}\right) = 0 \rightarrow N_j^C = \left(X_j^C / 2\right) \ \forall j, \tag{4}$$

$$\frac{\partial Z}{\partial N_k^D} = g^D \left(1 - \frac{2N_k^D}{X_k^D}\right) = 0 \rightarrow N_k^D = \left(X_k^D / 2\right) \ \forall k, \text{ and} \tag{5}$$

$$\frac{\partial Z}{\partial N_l^S} = g^S \left(1 - \frac{2N_l^S}{X_l^S}\right) = 0 \rightarrow N_l^S = \left(X_l^S / 2\right) \ \forall l. \tag{6}$$

These expressions state that each sector in an unregulated environment will produce at the peak of their appropriate logistic function—which is located halfway between 0 and *X*, as described in equations 3–6 and shown in Figure 1—without any consideration for abatement technologies. This will cause overproduction from an environmental perspective if agricultural production incurs a negative externality since firms will intensify to the point where marginal profit is equal to zero without accounting for the cost of pollution.

Nitrate emissions have socially-undesirable impacts off-farm in many catchments. Hence, regulatory bodies are becoming increasingly focused on restricting them. An appropriate regulatory or policy constraint may be stated:

$$N^T = \sum_{i=1}^{I} N_i^B + \sum_{j=1}^{J} N_j^C + \sum_{k=1}^{K} N_k^D + \sum_{l=1}^{L} N_l^S, \tag{7}$$

where N^T is the maximum level of nitrate emissions that the regulator will tolerate and is set exogenously. The terms on the Right Hand Side (RHS) of eq. 7 are total nitrate emissions arising from agriculture within the catchment. Nitrate targets may be defined in terms of proportional decreases in the unregulated level of leaching. This is achieved through the substitution of $N^T = (1-R)\left(\sum_{i=1}^{I} \frac{X_i^B}{2} + \sum_{j=1}^{J} \frac{X_j^C}{2} + \sum_{k=1}^{K} \frac{X_k^D}{2} + \sum_{l=1}^{L} \frac{X_l^S}{2}\right)$ into eq. 7, where *R*

[2] This results in a set of $i + j + k + l$ FOCs if profit functions are defined per-hectare.

is a proportional reduction in total nitrate leaching (e.g. $R = 0.1$ enforces a 10 per cent restriction). Eq. 7 is guaranteed to bind in any problem of practical relevance since there is a positive relationship between nitrate leaching and net returns on the left-hand side of the profit function (Figure 1).

The regulated catchment problem may be stated as maximisation of eq. 2 subject to eq. 7. This formulation is consistent with a market-based policy in which abatement is optimally allocated between sectors. Under optimal management, it will be typical for lower-abatement cost land uses to do more abatement per-hectare than those sectors with higher-abatement costs. This problem may be solved through formulating the Lagrangian function (Bazaraa et al., 2006):

$$L=\sum_{i=1}^{I} g^B N_i^B\left(1-\frac{N_i^B}{X_i^B}\right)+\sum_{j=1}^{J} g^C N_j^C\left(1-\frac{N_j^C}{X_j^C}\right)+\sum_{k=1}^{K} g^D N_k^D\left(1-\frac{N_k^D}{X_k^D}\right)+\sum_{l=1}^{L} g^S N_l^S\left(1-\frac{N_l^S}{X_l^S}\right)+ \lambda\left\{N^T-\left(\sum_{i=1}^{I} N_i^B+\sum_{j=1}^{J} N_j^C+\sum_{k=1}^{K} N_k^D+\sum_{l=1}^{L} N_l^S\right)\right\}, \quad (8)$$

where λ is a non-negative shadow price variable denoting the change in the optimal value of the objective function (defined in eq. 2) when the regulatory constraint (defined in eq. 7) is relaxed by a single unit. This function yields the FOCs:

$$\frac{\partial L}{\partial N_i^B} = g^B\left(1-\frac{2N_i^B}{X_i^B}\right)-\lambda = 0, \quad (9)$$

$$\frac{\partial L}{\partial N_j^C} = g^C\left(1-\frac{2N_j^C}{X_j^C}\right)-\lambda = 0, \quad (10)$$

$$\frac{\partial L}{\partial N_k^D} = g^D\left(1-\frac{2N_k^D}{X_k^D}\right)-\lambda = 0, \quad (11)$$

$$\frac{\partial L}{\partial N_l^S} = g^S\left(1-\frac{2N_l^S}{X_l^S}\right)-\lambda = 0, \text{ and} \quad (12)$$

$$\frac{\partial L}{\partial \lambda} = N^T-\left(\sum_{i=1}^{I} N_i^B+\sum_{j=1}^{J} N_j^C+\sum_{k=1}^{K} N_k^D+\sum_{l=1}^{L} N_l^S\right) = 0. \quad (13)$$

Eq. 9–12 are equivalent to conditions 3–6, except that environmental restrictions enter each expression through the (shared) shadow price variable. Eq. 13 is repetition of the policy constraint, as is standard when the Lagrangian function is differentiated with respect to the shadow price. The FOCs in equations 9–12 are both necessary and sufficient for a constrained maximum given the concavity of the objective function and the linear restraint (Bazaraa et al., 2006).

Equations 9–12 state that the shadow price is the marginal profit associated with nitrate emissions in each sector. Solving each of the conditions in eq. 9–12 for λ and setting them equal to one another yields:

$$g^B\left(1-\frac{2N_i^B}{X_i^B}\right)=g^C\left(1-\frac{2N_i^C}{X_i^C}\right)=g^D\left(1-\frac{2N_i^D}{X_i^D}\right)=g^S\left(1-\frac{2N_i^S}{X_i^S}\right). \tag{14}$$

This states that the rate at which profit changes with a marginal unit change in the policy constraint is equal between each sector in equilibrium. This highlights that an optimal solution requires coordination between all emitters and that all gains accruing to the trading of emission allowances between sectors are exhausted in equilibrium. Conditions 9–13 can be manipulated to obtain a closed-form solution for the shadow price variable. Alternatively, the global optimum can be identified in a straightforward manner using any NLP solver.

A common alternative to a market-based policy is a uniform emissions standard. This instrument requires each land use or firm to decrease their respective emissions by a fixed percentage that is constant across the catchment. This model may be stated as the maximisation of eq. 2 subject to multiple constraints:

$$(1-R)\sum_{i=1}^{I}\frac{X_i^B}{2}=\sum_{i=1}^{I}N_i^B, \tag{15}$$

$$(1-R)\sum_{j=1}^{J}\frac{X_j^C}{2}=\sum_{j=1}^{J}N_j^C, \tag{16}$$

$$(1-R)\sum_{k=1}^{K}\frac{X_k^D}{2}=\sum_{k=1}^{K}N_k^D, \text{ and} \tag{17}$$

$$(1-R)\sum_{l=1}^{L}\frac{X_l^S}{2}=\sum_{l=1}^{L}N_l^S, \tag{18}$$

where R is the fixed proportional reduction in emissions required in each land use. The specification of individual constraints for each land use makes the optimisation problem separable. That is, the optimal solution to the general problem can be identified through the collection of the optimal solutions to the individual optimisation problem faced by each sector. For example, the optimal level of intensity for the bull beef land use under a uniform policy is determined as the solution to: $\max_{N_i^B} Z_B=\sum_{i=1}^{I}g^B N_i^B\left(1-\frac{N_i^B}{X_i^B}\right)$, subject to

$(1-R)\sum_{i=1}^{I}\frac{X_i^B}{2}=\sum_{i=1}^{I}N_i^B$. Formulation of the Lagrangian for this problem leads to the FOCs:

$$\frac{\partial L}{\partial N_i^B}=g^B\left(1-\frac{2N_i^B}{X_i^B}\right)-\lambda^B=0\text{, and} \tag{19}$$

$$\frac{\partial L}{\partial \lambda^B}=(1-R)\sum_{i=1}^{I}\frac{X_i^B}{2}-\sum_{i=1}^{I}N_i^B=0. \tag{20}$$

This set of conditions derived for the bull beef land use, and those for the other sectors that closely resemble it, differs from the FOCs for the market-based instrument in a number of important ways. First, there are as many shadow prices as there are land uses, reflecting the presence of a single policy constraint for each sector. Second, the optimisation problem for each land use is solved separately; thus, an optimal solution does not require co-ordination between all parties. This is reflected in that generally $\lambda^B \neq \lambda^C \neq \lambda^D \neq \lambda^S$, in contrast to the equilibrium condition derived above for the market-based instrument. Finally, this problem may be solved efficiently in large applications by virtue of its separability.

3. Application

3.1. Parameter Values

The profit functions are estimated for the catchment using the following method. A large number of alternative production strategies for each land use are first generated, reflecting different farming intensities and mitigation technologies. Corresponding levels of profit and nitrate leaching are computed for each strategy using FARMAX® (FARMAX, 2009) and OVERSEER® (Monaghan et al., 2007) programmes, respectively. Model output is tailored to the Lake Taupo catchment through formulating appropriate assumptions regarding climate and production characteristics. In addition, leaching and profit are calculated for a forestry option, which had the lowest level of leaching at 3 kg N ha^{-1} yr^{-1}. Data sets for each sector are combined to allow decreases in nitrate leaching through retention of the current farming system and the adoption of mitigation strategies or a switch to less-intensive land uses (Table 1).

The profit curves are estimated using SHAZAM econometrics software (Whistler et al., 2004). The Cochrane-Orcutt method incorporating AR(1) errors (Cochrane and Orcutt, 1949) is used, as positive autocorrelation cannot be rejected at the 1 per cent significance level in all samples given the use of common models to generate data points. Each profit function is estimated using linear regression using the following analogue of eq. 1: $f(N)=gN-\left(\frac{g}{X}\right)N^2$, where $\beta_1=g$ and $\beta_2=\left(\frac{g}{X}\right)$ are the explanatory variables and

X is recovered through $X = \frac{\beta_1}{\beta_2} = \frac{g}{\beta_2}$. The results of these regressions are reported in Table 2. The number of data points is generally low. However, the estimated equations possess excellent explanatory power for a cross-sectional analysis, with the lowest adjusted R^2 being 0.62 for the sheep and beef farming data set.

Table 1. Data set for each sector used to estimate the logistic profit functions. Profit has been determined using FARMAX® software, while the level of nitrate leaching associated with these alternative strategies has been determined using OVERSEER® software

Activity	Nitrate leaching (kg N ha^{-1} yr^{-1})	Profit ($NZ ha^{-1} yr^{-1})	Data sets in which activity is used
Forestry	3	60	B, C, D, S[1]
Sheep and beef base farm	16	463	B, C, D, S
Sheep and beef farm, no winter nitrogen application	15	451	B, C, D, S
Sheep and beef farm, graze sheep on steep land and beef on rolling land plus nitrogen application	17	429	B, C, D, S
Sheep and beef farm, higher lambing percentage	16	512	B, C, D, S
Sheep farm, use of nitrification inhibitors	13	413	B, C, D, S
Sheep and deer farm, optimised scenario, 7 kg N ha^{-1} yr^{-1} applied	12	676	D, S
Sheep and deer farm, 42 kg N ha^{-1} yr^{-1} applied	14	687	D, S
Sheep and deer farm, 80 kg N ha^{-1} yr^{-1} applied	16	683	D, S
Sheep and deer farm, 160 kg N ha^{-1} yr^{-1} applied	22	640	D, S
Sheep and deer farm, higher breeding deer proportion, 160 kg N ha^{-1} yr^{-1} applied	22	445	D, S
Deer base farm, 7 kg N ha^{-1} yr^{-1}	11	1003	D
Deer farm, 42 kg N ha^{-1} yr^{-1}	12	1015	D
Deer farm, 80 kg N ha^{-1} yr^{-1}	14	1017	D
Deer farm, 160 kg N ha^{-1} yr^{-1}	20	990	D
Sheep and bull farm, 7 kg N ha^{-1} yr^{-1}	12	669	B, C, S[2]
Sheep and bull farm, 42 kg N ha^{-1} yr^{-1}	14	646	B, C, S
Sheep and bull farm, more bulls, 42 kg N ha^{-1} yr^{-1}	14	667	B, C, S
Sheep and bull farm, 80 kg N ha^{-1} yr^{-1}	17	637	B, C, S
Increase bulls only, 160 N sheep and bulls	23	611	B, C, S
Bull base farm, 7 kg N ha^{-1} yr^{-1}	11	613	B, C
Bull farm, 42 kg N ha^{-1} yr^{-1}	15	805	B, C
Bull farm, 80 kg N ha^{-1} yr^{-1}	18	856	B, C
Bull farm, 160 kg N ha^{-1} yr^{-1}	24	786	B, C
Base dairy farm	54	1614	C

Activity	Nitrate leaching (kg N ha^{-1} yr^{-1})	Profit ($NZ ha^{-1} yr^{-1})	Data sets in which activity is used
Dairy farm with 2% land reduction	52	1605	C
Dairy farm with 5% land reduction	50	1477	C
Dairy farm with DCD nitrification inhibitor and 6% more grass growth	47	1589	C
Dairy farm with 15 kg N ha^{-1} yr^{-1} on 1 August	52	1591	C
Dairy farm with alternative grazing system	54	1633	C
Dairy farm with halved N application and lower stocking rate	40	1670	C
Dairy farm with no N application, lower stocking rate, and later calving	29	1311	C
Organic dairy farm, all stock grazed on-farm and no imported feed	30	1030	C
Organic dairy farm, no imported feed but with young stock grazed off-farm all year and herd grazed off-farm for 6 weeks	24	1160	C
Dairy farm, reduced stocking rate but high production per cow with 20 days longer lactation and more pasture topping	51	1768	C
Dairy farm, all cows wintered off for 6 weeks, increased stocking rate, calving 10 days earlier	52	1793	C
Dairy farm, all cows wintered in herd home for 6 weeks, increased stocking rate, calving 10 days earlier	52	1490	C
Dairy farm, all cows wintered on feed pad for 6 weeks, increased stocking rate, calving 10 days earlier	52	1604	C
Dairy farm, all cows wintered on feed pad for 18 hours a day, increased stocking rate, calving 10 days earlier	50	1557	C
Dairy farm, no N application, replace nitrogen-boosted grass with maize concentrate	33	1485	C
Dairy farm, no N application, all cows wintered off	28	1608	C
Dairy farm, no N application, all cows wintered in herd home	32	1323	C
Dairy farm, no N application, all cows wintered on feed pad	32	1428	C

[1] B=bull beef, C=dairy, D=deer, and S=sheep and beef farming.

[2] Deer options are not included in the bull beef, dairy, and sheep farming scenarios due to the high switching costs associated with establishing these farming systems.

Table 2. Regression results for parameters of the logistic profit functions derived for each sector

Description	Sector			
Bull	Dairy	Deer	Sheep	
Observations	15	34	15	16
Adjusted R^2	0.78	0.89	0.72	0.62
β_1 coefficient	70.74	50.61	108.53	74.83
β_1 p-value	0.000	0.000	0.001	0.000
β_1 95% conf. interval	[45.14, 96.34]	[41.64, 59.58]	[59.77, 157.29]	[58.38, 91.29]
β_2 coefficient	-1.97	-0.38	-4.04	-2.34
β_2 p-value	0.001	0.000	0.001	0.000
β_2 95% conf. interval	[-2.9, -1.03]	[-0.56, -0.2]	[-5.87, -2.2]	[-3.1, -1.57]
g	70.74	50.61	108.53	74.83
X	35.99	134.67	26.89	32.05

Land area for each sector is drawn from a variety of sources. The total area of agricultural land is 46,079 ha. This consists of 42,306 (92 per cent) of sheep and beef farming; 1,931 ha (4 per cent) of dairy farming; and 922 ha (2 per cent) each of deer and bull beef farming. The total area of agriculture and the area of dairy farming in the catchment are obtained from Environment Waikato (unpublished data). The proportion of land used for bull beef and deer farming is based on the New Zealand average (Ministry of Agriculture and Forestry, 2009). The remainder is allocated to sheep and beef farming, which dominates the other sectors in terms of land allocation in the catchment.

3.2. Model Scenarios

The base case (Section 4.1) discusses the abatement cost curves for each sector and the change in profit associated with nitrate regulations enforcing 0–100 per cent reductions using the standard parameter values in the model.

The impact of alternative land allocations on abatement costs are investigated in Section 4.2. The land allocations employed denote percentages of land allocated to bull-dairy-deer-sheep and beef farms. The five scenarios are: (a) 2-4-2-92 (base case), (b) 7.5-7.5-15-70, (c) 10-10-10-70, (d) 15-15-15-55, and (e) 25-25-25-25.

The abatement-cost curve derived for the base case is compared with alternative formulations computed for different specifications of the profit curve for sheep and beef farms in Section 4.3. Endpoints for the 95 per cent confidence intervals for the parameters of this profit curve (Table 2) are used to derive nine alternate cases (Table 3). These scenarios are generated through identification of low ($g = 58.38$), medium ($g = 74.83$), and high ($g = 91.29$) estimates of g and derivation of corresponding levels of X through the relationship $X(g) = \beta_1 \left|\beta_2^{-1}\right|$ for $\beta_1 = g$ and $\beta_2 = \{-3.1, -2.34, -1.57\}$.

Table 3. Alternative profit function parameters for sheep and beef farming investigated in sensitivity analysis

Scenario	Low slope ($g = 58.38$)	Scenario	Medium slope ($g = 74.83$)	Scenario	High slope ($g = 91.29$)
Case I	$X = 18.83$	Case IV	$X = 24.14$	Case VII	$X = 29.44$
Case II	$X = 25.14$	Case V (base case)	$X = 32.04$	Case VIII	$X = 39.01$
Case III	$X = 37.18$	Case VI	$X = 47.66$	Case IX	$X = 58.15$

Section 4.4 investigates the relative value of market-based and uniform policy instruments. The impact of different land allocation scenarios on abatement cost is investigated through subtraction of the abatement costs derived for a market-based policy from those computed for a uniform policy for a number of cases. The following land-allocations scenarios are discussed for bull-dairy-deer-sheep and beef farms. The cases are: (1) 2-4-2-92 (base case), (2) 8.3-75-8.3-8.3 (75 per cent dairy), (3) 2-92-2-4 (92 per cent dairy), and (4) 25-25-25-25 (even allocation).

The model is solved with NLP using the CONOPT3 solver in GAMS Distribution 23.0 (Brooke et al., 2008). The program is available from the primary author on request.

4. Results and Discussion

4.1. Base Scenario

The cost curves for bull beef, deer, and sheep and beef production are quite uniform, though that for deer is relatively steep (Figure 2). The cost curve for dairy is broadly disparate, but shows that significant absolute reductions in nitrate leaching can be achieved on these farms at relatively low cost given the available mix of mitigation technologies. Nonetheless, there is little scope for large leaching reductions at the catchment scale through greater regulation of dairy farming since only 4 per cent of the catchment consists of this form of agriculture. Moreover, the high nitrate leaching burden in absolute terms imposed by dairy production (Table 1) precludes the expansion of this sector under regulation.

Low-to-moderate reductions in leaching can be achieved at relatively low cost on sheep and beef farms given the available mitigation technologies (Figure 2). For example, the use of nitrification inhibitors (Monaghan et al., 2007) decreases leaching by 19 per cent, but reduces profit by only 4 per cent (Table 1). Moreover, these systems dominate the catchment in terms of land area; thus, reasonable reductions in nitrate leaching in the catchment do not bear a large cost. However, profits decrease at an increasing rate, as more stringent nitrate restrictions require greater changes to underlying production systems within each sector (Figure 3). For example, sheep and beef producers must bear a substantial cost as an increasing proportion of their land is used for forestry as emissions are regulated more heavily.

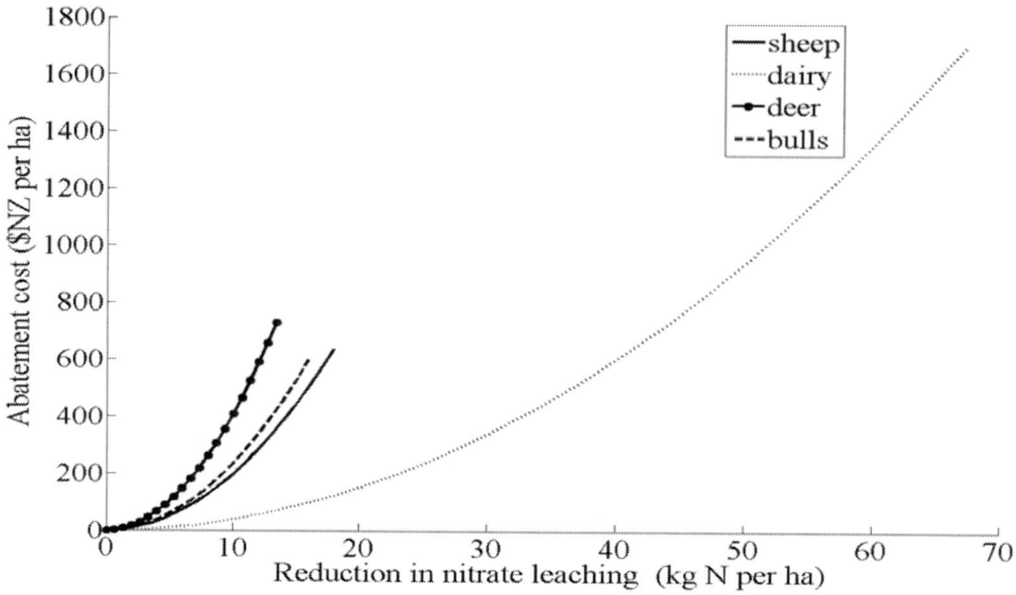

Figure 2. Abatement cost functions for bull beef production, dairy production, deer production, and sheep and beef production generated from optimisation models for each sector.

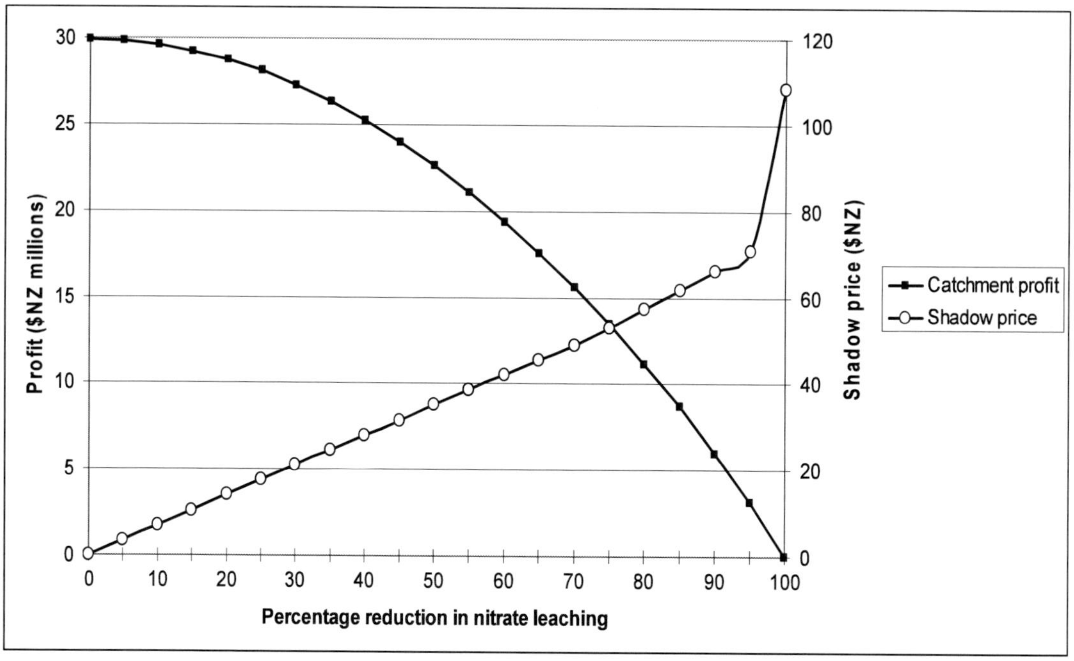

Figure 3. Catchment profit and shadow price for percentage reductions in nitrate leaching.

The shadow price of the policy constraint is the change in optimal profit associated with a 1 kg increase in the level of nitrate restriction imposed on the catchment. Theoretically, it is the price for the supply of nitrate allowances at which the market for tradable permits will clear. In addition, it is the optimal level of taxation at which the emissions level consistent with the binding emissions constraint will be attained. The model predicts that a 10 (20) per cent reduction in base emissions will be achieved at a price/tax of around $NZ7/kg N

($NZ14/kg N) (Figure 3). This marginal benefit increases with the stringency of regulation, as the value of foregone production is higher when farming is constrained to a greater degree.

The current policy involves farm emissions being capped at their highest recorded level over 2004–07 and the Lake Taupo Protection Trust aiming to decrease leaching by 20 per cent. A least-cost strategy involves reducing emissions from sources with the lowest abatement cost. The model estimates that this imposes an annual cost of $NZ1,169,610 or 4 per cent of total profit (Figure 3).[3] This reinforces that the optimal adoption of available abatement technologies in response to emission restrictions can greatly reduce those losses being incurred at low-to-moderate levels of nitrate regulation. However, these minimal losses will only be borne if producers adopt appropriate mitigation technologies, highlighting an important role for agricultural extension. Indeed, costs will be magnified for some time in reality, as producers trial and adopt different technologies in response to the imposition of regulatory policy.

Dairy farms are regulated more stringently as the intensity of regulation increases (Figure 4) since abatement costs per unit of N are initially lower in this sector and its environmental impact is high in relative terms (Figure 2). Indeed, no dairy production occurs with more than a 45 per cent reduction in leaching because it is necessary to shift to bull beef production to sufficiently reduce emissions.[4] In contrast, deer production is intensified given its greater nitrogen-use efficiency.

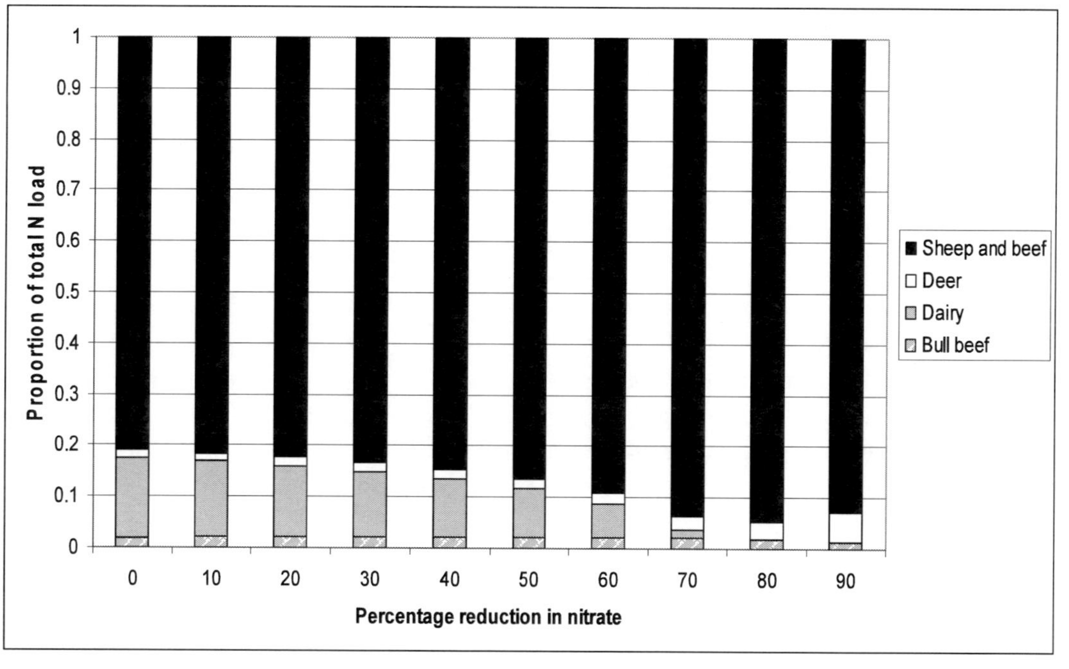

Figure 4. The proportion of total leaching load within each land use for different levels of nitrate reduction enforced using a differentiated policy at the standard land allocation.

[3] This may be perceived as a one-off cost; however, an annual value is relevant since the policy incurs an opportunity cost each year, equal to the value of foregone production.

[4] This is ascertained through matching nitrate leaching in the dairy sector at this level of regulation with the corresponding mitigation strategy in Table 1.

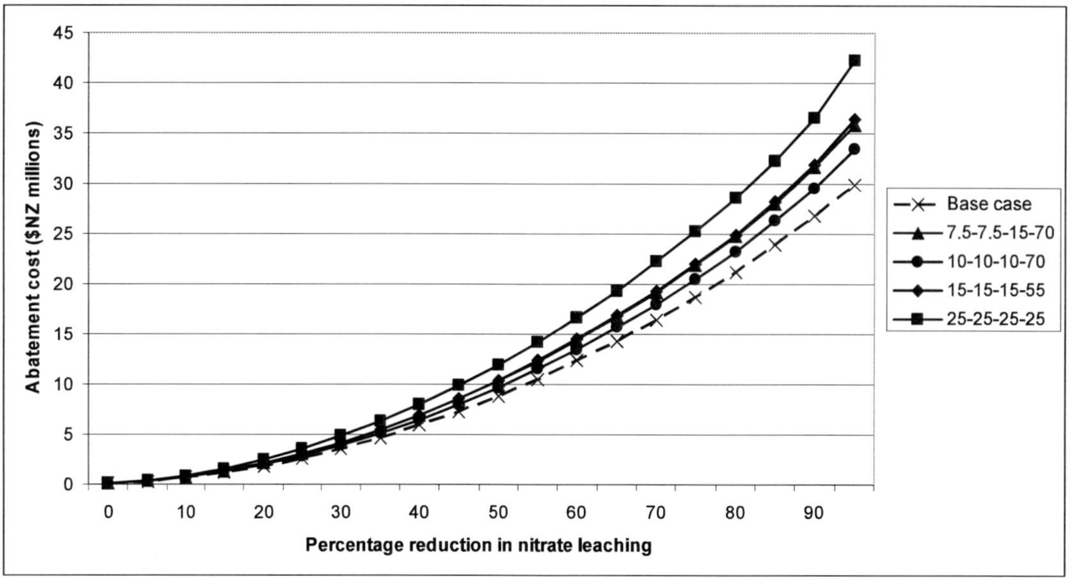

Figure 5. Abatement cost curves for alternative scenarios incorporating different land use allocations within the Lake Taupo catchment. Figures in the legend for each scenario designate the percentage of total land use allocated to bull beef-dairy-deer-sheep and beef farming.

4.2. Sensitivity of Abatement Cost to Different Land Allocations

Abatement cost increases as the proportion of sheep and beef farming is reduced (Figure 5), as this impacts the overall availability of cost-effective abatement technologies. The highest abatement cost at elevated levels of nitrogen regulation is that derived for equal land use allocation (i.e. the "25-25-25-25" scenario) (Figure 5). This identifies the significant cost associated with high abatement in the simulated dairy sector. Also, the "7.5-7.5-15-70" and "15-15-15-55" cases are closely equivalent, as the 15 per cent reduction in sheep and beef farming in the latter scenario is offset with a collective 15 per cent increase in bull beef and deer farming, and these three land uses possess relatively uniform abatement cost curves, particularly in relation to dairy farming (Figure 2). However, overall, there is little difference in abatement cost at low-to-moderate levels of regulation when land allocation for each enterprise is altered (Figure 5). Model output is therefore robust to large errors surrounding initial land allocation.

4.3. Sensitivity of Abatement Cost to Different Profit Relationships

Sheep and beef farming dominates the study region. Thus, broad changes in its profit function substantially alter the abatement-cost profile derived for the catchment. Figure 6 presents abatement cost curves derived for slopes of the profit curve in this sector of low ($g = 58.38$) (Fig. 6a), medium ($g = 74.83$) (Fig. 6b), and high ($g = 91.29$) (Fig. 6c) magnitudes. The abatement cost never (always) surpasses that of the base case if the emissions rate is fixed at its simulated lower (higher) bound (cf. Fig. 6a and 6c). Moreover, abatement cost increases monotonically with the maximum level of nitrate leaching (X) described by a given profit curve (Figures 6a–6c); this reflects that unconstrained production occurs at $0.5X$ in each scenario (see eq. 3–6) and abatement cost increases with augmentation of the profit function.

The broad range of the coefficients obtained from the regression procedure yield quite dissimilar abatement-cost profiles at high percentage reductions in nitrate leaching. However, the cost estimates corresponding to lower, more appropriate targets for regulation are more uniform. For example, even if X takes its highest value for each value of g simulated, the cost of 20 per cent regulation is only around $NZ1.1 million, $NZ1.69 million, and $NZ2.43 million for the low, medium, and high emissions rate scenarios, respectively. Thus, the cost of low-to-moderate emissions reductions is limited if abatement strategies are fully adopted by farmers, even if major errors surround the estimation of the profit-leaching relationship for sheep and beef farming.

4.4. Benefits of a Market-based versus Uniform Instrument

A market-based policy is more cost-effective than a uniform policy (Figure 7), as market-based instruments allow the allocation of emission entitlements according to abatement efficiency. Nonetheless, the difference between the curves derived for each policy is very low (Figure 7), with abatement cost being, on average, around only 2 per cent lower for the market-based policy. These cost savings are at the lower range of what could be expected, equivalent to the savings identified by Helfand and House (1995).

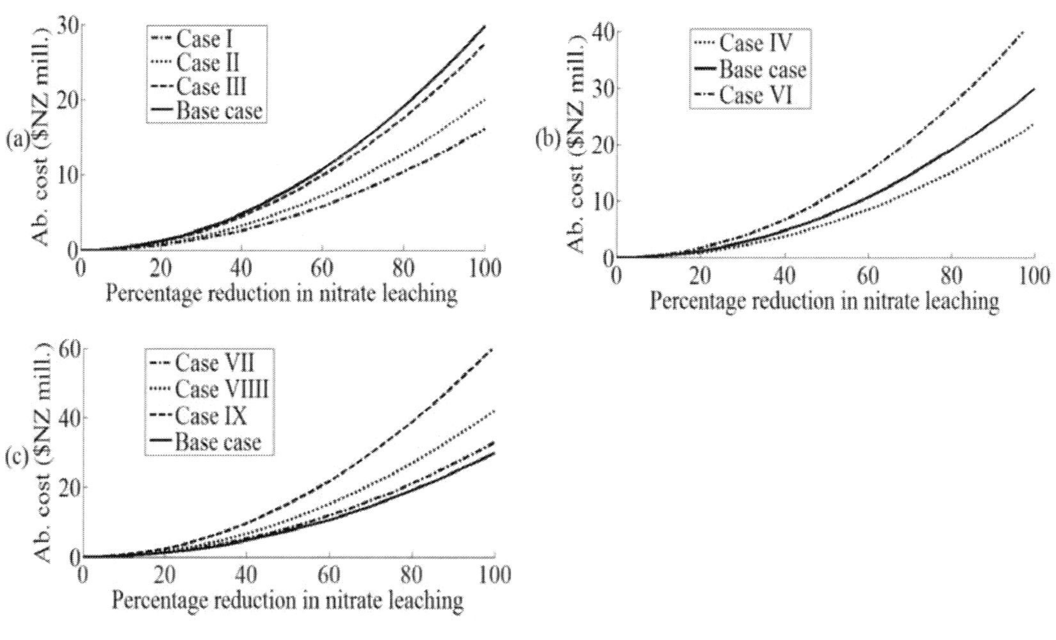

Figure 6. Sensitivity of the abatement-cost function to perturbations in the curvature and intercept of the profit function estimated for the sheep and beef farming sector. Figures present cost curves for (a) Cases 1–3 and the base case, (b) Cases 4–6, and (c) Cases 7–9 and the base case.

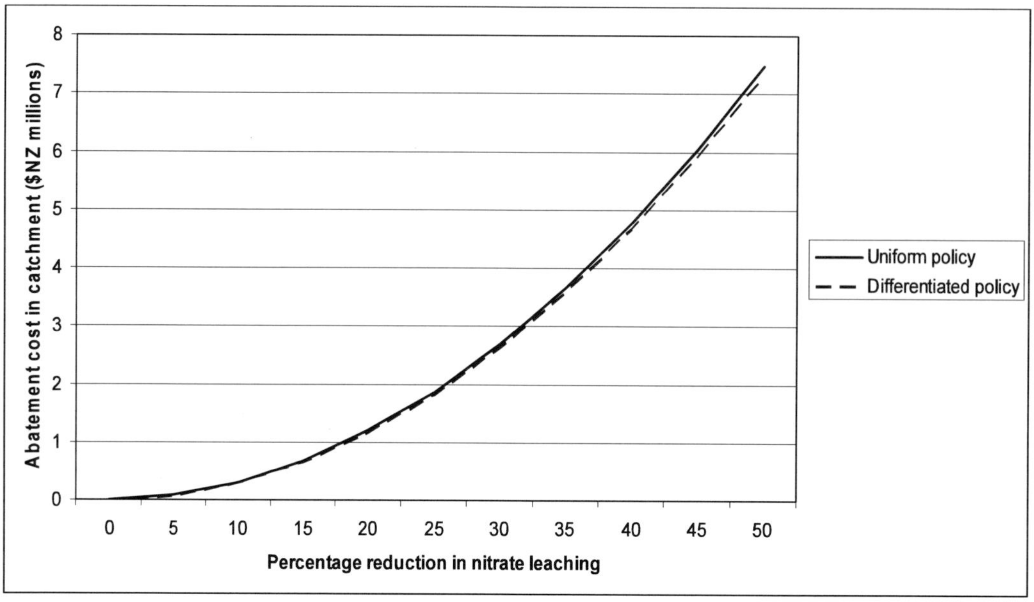

Figure 7. Abatement cost functions for nitrate leaching regulations enforced through a uniform percentage reduction in each sector or through a differentiated policy involving optimisation of the relative contributions of each land use.

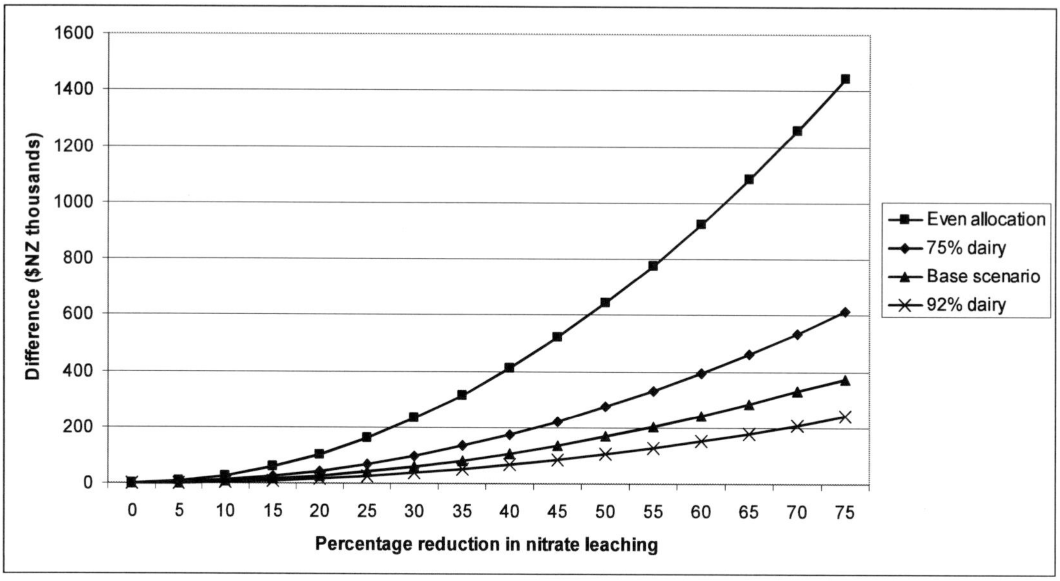

Figure 8. The difference in cost between a differentiated and uniform policy instrument for four scenarios: (a) 2-4-2-92 (base case), (b) 8.3-75-8.3-8.3 (75 per cent dairy), (c) 2-92-2-4 (92 per cent dairy), and (d) 25-25-25-25 (even allocation).

The relative benefits accruing to a market-based policy are even lower if 92 per cent of the catchment consists of dairy farming, instead of sheep and beef farming (Figure 8). This demonstrates that the domination of the catchment by a single land use reduces the economic benefit of a market-based policy, regardless of its intensity. Indeed, Figure 8 shows that these benefits are further eroded if the abatement-cost profile for the dominant form of agriculture (dairy farming in this scenario) is lower than that of the competing land uses. In comparison, the benefit of a market-based policy nearly doubles if three-quarters of the catchment contains dairy farming. Moreover, relatively large gains are associated with a market-based instrument if land allocation in the catchment is divided evenly between the four sectors. In fact, they are more than twice that of the 75 per cent dairy scenario and nearly four times greater than that for the base case.

Firm and sector homogeneity determine the value of a market-based policy, relative to a uniform instrument (Newell and Stavins, 2003). The former is not studied here due to data limitations. Thus, increased homogeneity is present, consistent with the large majority of economic models of environmental policy regulation (e.g. Nordblom et al., 2006) in which individual farms are not represented. Nonetheless, firm homogeneity is more likely in bull beef, deer, and sheep and beef farming sectors given the low range of potential abatement that producers may achieve in these systems in absolute terms (Figure 3). In addition, Figure 8 demonstrates that land use homogeneity is a key determinant undermining the superiority of a market-based policy. This is an important finding since agricultural catchments throughout Australasia, particularly New Zealand, are typically dominated by certain land uses, as these watersheds are often characterised by a low variance in characteristics that determine the comparative advantage of a given enterprise. Examples of important characteristics are climate, labour and capital supply, pests and disease, rainfall intensity and distribution, soil type, and topography.

CONCLUSION

Uncertainty complicates the assessment of NPS policies. Accordingly, this paper presents a general framework for the evaluation of policies for NPS pollution mitigation in agricultural catchments. This model overcomes the need for the identification of relevant stochastic processes by relying on structured sensitivity analysis to guide the development of understanding regarding a given problem. The model is used to estimate the abatement costs corresponding to existing and prospective policies in the Lake Taupo catchment and determine the relative value of market-based and uniform policy instruments.

Model output demonstrates that up to 30 per cent reductions in unregulated nitrate emissions can be achieved without incurring substantial abatement costs. Dairy production is regulated more heavily than other sectors initially, as this sector has a low abatement cost per unit of emissions at lower levels of regulation. Moreover, a large proportion (over 90 per cent) of the catchment consists of sheep and beef farming, in which there exists cost-effective means to achieve low-to-moderate leaching reductions. This finding is robust to substantial changes in the assumed land use allocation and the parameters defining the relationship between profits and leaching load within the sheep and beef farming sector. Low abatement costs will only be borne if available mitigation technologies are fully adopted by landholders. Thus, practice change should be critically supported by agricultural extension and research into improving the targeting and implementation of mitigation practices. Furthermore, investment in research and development for new technologies or improving the integration of available/new innovations could further reduce abatement costs.

A market-based policy incurs a cost only marginally below that associated with a uniform instrument, and any differences are so small that they are likely to be subsumed by transaction cost reductions. Thus, uniform reductions can be a cost-effective method, relative to market-based instruments, to achieve water quality targets in Australasian catchments dominated by a single form of agriculture. This is an important finding since the use of market-based policies for regulating agricultural emissions is increasing, both in Australasia (Lock and Kerr, 2008) and worldwide (De Cara et al., 2005), and their implementation is complicated by (1) the complexity and cost of initial allocation; (2) high transaction costs, especially for establishment and trading; and (3) thin permit markets. However, this conclusion is dependent on the level of firm heterogeneity present within the dominant sector. Thus, future research investigating its implications for the relative worth of market-based and uniform instruments in agricultural catchments is worthwhile.

REFERENCES

Bazaraa, M.S, Sherali, H.D., and Shetty, C.M. (2006), *Nonlinear programming: theory and algorithms*, Wiley, New York.

Brooke, A., Kendrick, D., Meeraus, A., and Raman, R. (2008), *GAMS—A user's guide*, GAMS Development Corporation, Washington D. C.

Cochrane, D. and Orcutt, G. H. (1949), 'Applications of least square regression to relationships containing autocorrelated error terms', *Journal of the American Statistical Association* 44, pp. 32–61.

De Cara, S., Houze, B.M., and Jayet, P.A. (2005), 'Methane and nitrous oxide emissions from agriculture in the EU: A spatial assessment of sources and abatement costs', *Environmental and Resource Economics* 32, pp. 551–583.

Doole, G.J. (2008), 'Optimal management of annual ryegrass (*Lolium rigidum* Gaud.) in lucerne-wheat rotations in the Western Australian wheatbelt', *Australian Journal of Agricultural and Resource Economics* 52, pp. 339–362.

Doole, G. J. and Kingwell, R. S. (2010), 'Robust mathematical programming for natural resource modeling under parametric uncertainty', *Natural Resource Modelling* 23, pp. 283–302.

Elliot, A.H. and Stroud, M.J. (2001), *Prediction of nutrient loads entering Lake Taupo under various land use scenarios*, NIWA Client Report EVW01224, Hamilton.

FARMAX (2009), *FARMAX agricultural decision making software*. [Online]. URL: http://*www.farmax.co.nz*. [Accessed: 6 December 2008]

Gibbs, M. (2008), *Lake Taupo long-term monitoring programme 2006-2007*, Environment Waikato Technical Report 2008/25, Hamilton.

Hadfield, J. (2007), *Groundwater flow and nitrogen transport modelling of the northern Lake Taupo catchment*, Environment Waikato Technical Report 2007/39, Hamilton.

Helfand, G.E. and House, B.W. (1995), 'Regulating nonpoint source pollution under heterogeneous conditions', *American Journal of Agricultural Economics* 77, pp. 1024–1032.

Lock, K. and Kerr, S. (2008), *Nitrogen trading in Lake Rotorua: choosing the scope of a nutrient trading scheme*, Motu Working Paper 08-05, Wellington.

Ministry of Agriculture and Forestry (2009), *Land use in New Zealand*, Ministry of Agriculture and Forestry, Wellington. [Online]. URL: http://www.maf.govt.nz/statistics/land-use/. [Accessed: 29 December 2009]

Monaghan, R.M., Hedley, M.J., Di, H.J., McDowell, R.W., Cameron, K.C., and Ledgard, S.F. (2007), 'Nutrient management in New Zealand pastures—recent developments and future issues', *New Zealand Journal of Agricultural Research* 50, pp. 181–201.

Newell, R.G. and Stavins, R.N. (2003), 'Cost heterogeneity and the potential savings from market-based policies', *Journal of Regulatory Economics* 23, pp. 43–59.

Nordblom, T., Hume, I., Bathgate, A.D., and Reynolds, M. (2006), 'Mathematical optimisation of drainage and economic land use for target water and salt yields', *Australian Journal of Agricultural and Resource Economics* 50, pp. 381–402.

Segerson, K. (1988), 'Uncertainty and incentives for nonpoint pollution control', *Journal of Environmental Economics and Management* 15, pp. 87–98.

Shortle, J.S. and Horan, R.D. (2001), 'The economics of nonpoint pollution control', *Journal of Economic Surveys* 15, pp. 255–289.

Suter, J.F., Vossler, C.A., Poe, G.L., and Segerson, K. (2008), 'Experiments on damage-based ambient taxes for nonpoint source polluters', *American Journal of Agricultural Economics* 90, pp. 86–102.

Vant, B., and Huser, B. (2000), 'Effects of intensifying catchment land use on the water quality of Lake Taupo', *Proceedings of the New Zealand Society of Animal Production* 6, pp. 261–264.

Weersink, A., Livernois, J., Shogren, J.F., and Shortle, J.S. 'Economic instruments and environmental policy in agriculture', *Canadian Public Policy*, Vol. 24, (1998) pp. 309–327.

Whistler, D., White, K.J., Wong, S.D., and Bates, D. (2004), *SHAZAM version 10 user's manual*, Northwest Econometrics, Gibsons.

In: Economics of Innovation, Incentives and Uncertainty
Editors: P. Simmons and S. T. Jordan
ISBN: 978-1-62257-251-9

Chapter 4

A USE OF DEA-DA FOR CORPORATE VALUE ASSESSMENT: A BUSINESS IMPLICATION OF R&D POLICY AND INVESTMENT

***Toshiyuki Sueyoshi*[1,*] *and Mika Goto*[2,†]**
[1]New Mexico Institute of Mining and Technology, Department of Management, Socorro, NM, US
[2]Central Research Institute of Electric Power Industry, Otemachi, Chiyodaku, Tokyo, Japan

ABSTRACT

This study discusses a use of DEA-DA (Data Envelopment Analysis-Discriminant Analysis) to assess the corporate value of firms. It is widely known among finance researchers that Tobin's q is a ratio to measure the corporate value. A contribution of this study can be found in a linkage between DEA-DA (as a methodology) and Tobin's q (as a theoretical basis). As an application of the combined use between DEA-DA and Tobin's q, this study investigates the influence of R&D (Research & Development) expenditure on the performance of Japanese manufacturing industries. This research identifies that the R&D expenditure is important for the Japanese manufacturing industries to be competitive in a global market. However, the importance of R&D expenditure depends upon the type of an industry. Empirical findings identified in this study are useful to other industries (not only manufacturing business) and other nations (not only Japan).

Keywords: R&D Policy, Technology Innovation, Finance

---

* E-mail: toshi@nmt.edu.
† E-mail: mika@criepi.denken.or.jp

INTRODUCTION

Discriminant Analysis (DA) is a classification method that can predict the group membership of a newly sampled observation. In a use of DA, a group of observations whose memberships are already identified are used to estimate the weights (or parameters) of a discriminant function by some criteria such as the minimization of misclassifications or the maximization of correct classifications. A new sample is classified into one of several groups by DA results.

Recently, Sueyoshi (2004, 2005 and 2006) has proposed a new type of nonparametric DA approach that provides a set of weights of a linear discriminant function(s), consequently yielding an evaluation score(s) for the determination of group membership. The new nonparametric DA is referred to as "Data Envelopment Analysis-Discriminant Analysis (DEA-DA)," because it maintains a discriminant capability by incorporating the nonparametric feature of DEA into DA.

After reviewing the previous research works, this study needs to mention that they have insufficiently explored a use of DEA-DA from the practical perspectives of financial evaluation. Hence, this study discusses the use of DEA-DA for the financial evaluation, focusing upon its linkage with the corporate value assessment based upon Tobin's q. No research has investigated the business linkage in the previous studies.

An enhancement in financial performance is the final goal of corporate activities, including technology innovation. In this study, a corporate effort for technology innovation is measured by the amount of R&D expenditure because Japanese corporate leaders believe that their firms can maintain a leading position in a global market as long as their technology outperforms that of the other firms. They believe that technology innovation is the most important strategy to compete in a global market. Such a business view was histrocally true, but it may be not true anymore. To examine the assertion, this study empirically investigates the view of Japanese manufacturing firms. That is the purpose of this study.

The remaining structure of this article is organized as follows: Section 2 includes a literature review that indicates the position of this research among existing literatures on DA. Section 3 describes the formulation of DEA-DA. Section 4 review the previous studies on Tobin's q. Section 5 applies the proposed DEA-DA to examine the corporate performance of Japanese manufacturing industries. The last section (6) concludes this study along with future research agendas.

LITERATURE REVIEW

The DA has the following research groups from its methodological developments and practical uses:

Statistics: A group of research is interested in statistical DA methods. The first contribution may be dating back to Fisher (1936) and Smith (1947). [See, for instance, Maddala (1983) and McLachlan (1992) which compiled previous contributions concerning statistical DA.] The conventional statistical DA methods usually maintain underlying assumptions on group distributions. For example, two groups come from a normal distribution with different means, but a same covariance matrix, all of which should be known to us before these applications. Under these assumptions, the statistical DA methods provide a theoretical basis for conducting various statistical inferences and tests. Furthermore, an ordinary least squares regression is usually used to obtain coefficient estimates of a linear discriminant function. Thus, there is a computational simplicity in the statistical DA methods. These are methodological strengths and a contribution, indeed. However, it is also true that many real data sets do not satisfy such underlying assumptions.

Econometrics: If independent variables are normally distributed, the statistical DA estimator is a true maximum-likelihood estimator and therefore, it is asymptotically more efficient than the other DA methods. However, the assumption on normality is not satisfied in many real data sets. To overcome such a shortcoming related to statistical DA, econometricians developed other several DA methods that had a close linkage with the theory of a probabilistic choice discussed by psychologists. The most well known research effort in this area was explored by McFadden (1980) who investigated logit and probit models. The two models can be solved by maximum-likelihood methods. An important feature of logit and probit analyses is that the two models provide the conditional probability of an observation belonging to a certain class, given its independent variables. Both are based on a cumulative probability function and neither requires that independent variables should be multivariate normal or that groups have equal covariance matrices, unlike the requirements of statistical DA. Furthermore, these approaches have a close linkage with statistical inferences and various tests.

Mathematical Programming: Mathematical Programming (MP) methods were proposed for solving various DA problems. These methods consisted of the third group in this study on DA methodology. The first contribution of this group was due to Charnes et al. (1955) who documented how to formulate L_1 metric regression by a goal programming model and how to solve the problem by linear programming algorithm. [See Charnes and Cooper (1977) for their description on goal programming.] A popularity of MP-based DA among researchers occurred after the research effort of Freed and Glover (1981a,b). They presented how a DA problem was formulated by goal programming. Based upon the optimization techniques to be used, the third group of DA studies was further classified into (a) linear programming methods (e.g., Mangasarian, 1999), (b) nonlinear programming methods (e.g., Duarte Silva and Stam, 1994, and Falk and Karlov, 2001) and (c) mixed integer programming methods (e.g., Bajgier and Hill, 1982 and Yanev and Balev, 1999). A comprehensive review on the MP-based DA was found in Doumpos et al. (2001) and Zopounidis and Doumpos (2002). A methodological benefit of this research group is that the MP-based DA methods do not need any assumption on a group distribution. Meanwhile, a shortcoming of the MP-based DA is that statistical inferences and tests are not yet well developed at the level of the statistical and econometric DA approaches.

Position of this research: It is clear that this study belongs to the third research group of DA. The previous DA studies in this research group have been looking for the development of new DA models and these related computational issues. DEA-DA was developed under such a research direction. Unfortunately, the previous research efforts on DEA-DA have not yet clearly discussed how to use it for real business applications. Hence, this study describes DEA-DA from a practical perspective of corporate assessment.

DEA-Discriminant Analysis

DEA-DA has two stages for computation. The first stage identifies an existence of an overlap between two groups of observations and it classifies observations not belonging to the overlap. The second stage classifies the observations in the overlap.

Stage 1: To discuss the first stage, we consider the following formulation:

$$\min \left\{ s \;\middle|\; \begin{array}{l} \sum_{f=1}^{h} \lambda_f z_{fj} - d + s \geq 0, j \in G_1, \sum_{f=1}^{h} \lambda_f z_{fj} - d - s \leq -\varepsilon, j \in G_2, \\ \sum_{f=1}^{h} |\lambda_f| = 1, \ \lambda_f, d \ \& \ s: \text{URS}. \end{array} \right\}. \tag{1}$$

The objective function of Model (1) minimizes an unknown variable (s) that indicates the size of an overlap between G_1 and G_2. The overlap is surrounded by d-s (as a lower bound of G_1) and d+s (as an upper bound of G_2). The discriminant score for group classification is expressed by a scalar value "d". Both d and s are unrestricted (URS). A very small number (ε) is incorporated into Model (1) in order to separate the two groups clearly. When an observation exists on an estimated discriminant function, we have a difficulty in its classification. The small number is incoporated in (1) for such classification convenience. All the factors (z_{fj}) of the j-th observation are connected by a discriminant function ($\sum_{f=1}^{h} \lambda_f z_{fj}$). The subscript (f) stands for the f-th financial factor. Furthermore, these weights are restricted in such a manner that the sum of absolute values of λ_f (for all f = 1, .., h) is unity. Consequently, Model (1) estimates each weight by a percentile expression, so that we can examine which weight is important or not in terms of group classification. The weight restriction is known as "normalization."

Sign of the unknown variable (s): The first stage identifies the existence of an overlap by $s^* \geq 0$ on optimality of Model (1). An opposite case ($s^* < 0$) indicates no overlap. If the first stage incorporates the constraint of non-negativity on the variable (s) in Model (1), then $s^* = 0$ occurs on the following two cases. One of the two cases is that the variable becomes zero because Model (1) can classify two groups, but there is a single intersection between them. In this sense, $s^* = 0$ implies an occurrence of overlap. The other case is that the variable becomes zero because Model (1) incorporates the non-negativity even if $s^* = 0$ implies no overlap. The variable should be negative in the case of no restriction. To distinguish between the two cases, the variable (s) should be unrestricted in (1).

Model (1) has the following reformulation because it is impossible for us to solve Model (1) directly.

$$\min\left\{ s \left| \begin{array}{l} \sum_{f=1}^{h} (\lambda_f^+ - \lambda_f^-) z_{fj} - d + s \geq 0, j \in G_1, \sum_{f=1}^{h} (\lambda_f^+ - \lambda_f^-) z_{fj} - d - s \leq -\varepsilon, j \in G_2, \\ \sum_{f=1}^{h} \left(\lambda_f^+ + \lambda_f^- \right) = 1, \zeta_f^+ \geq \lambda_f^+ \geq \varepsilon \zeta_f^+ \text{ (for all f)}, \zeta_f^- \geq \lambda_f^- \geq \varepsilon \zeta_f^- \text{ (for all f)}, \\ \zeta_f^+ + \zeta_f^- \leq 1 \text{ (for all f)}, \ \lambda_f^+ + \lambda_f^- \geq \varepsilon \ \text{ (for all f)}, \\ d \,\&\, s: \text{URS}, \zeta_f^+ \,\&\, \zeta_f^- : \text{binary, and all other variables} \geq 0. \end{array} \right. \right\}, \tag{2}$$

In transforming λ_f (f = 1,..,h) into a special ordered set of paired variables ($\lambda_f = \lambda_f^+ - \lambda_f^-$), we define these variables as

$$\lambda_f^+ = \left(|\lambda_f| + \lambda_f\right)/2 \text{ and } \lambda_f^- = \left(|\lambda_f| - \lambda_f\right)/2, \tag{3}$$

each representing positive or negative part of λ_f, respectively. These paired variables maintain the following relationships: $\lambda_f = \lambda_f^+ - \lambda_f^-$ and $|\lambda_f| = \lambda_f^+ + \lambda_f^-$.

Such a transformation needs Non-Linear Condition (NLC: $\lambda_f^+ \lambda_f^- = 0$) for each f (= 1,…,h) in order to avoid a simultaneous occurrence of both $\lambda_f^+ > 0$ and $\lambda_f^- > 0$. To incorporate NLC, this study uses its MIP (Mixed Integer Programming) equivalence. Here, let ζ_f^+ (= 0 or 1) and ζ_f^- (= 0 or 1) be two binary variables, then NLC is expressed by

$$NLC: \ \zeta_f^+ \geq \lambda_f^+ \geq \varepsilon \zeta_f^+ \text{ and } \zeta_f^- \geq \lambda_f^- \geq \varepsilon \zeta_f^- \quad (f = 1, .., h) \tag{4}$$

$$\zeta_f^+ + \zeta_f^- \leq 1 \ (f = 1, .., h) \tag{5}$$

Here, Equation (4) implies the upper and lower bounds of λ_f^+ and those bounds of λ_f^-, respectively. Furthermore, Equation (5) implies that the sum of these binary variables is less than or equal to unity. If both $\lambda_f^+ \geq \varepsilon > 0$ and $\lambda_f^- \geq \varepsilon > 0$ occur simultaneously in Equation (4), then Equation (5) becomes $\zeta_f^+ + \zeta_f^- = 2$. The result is infeasible in Equation (5). Hence, the simultaneous occurrence of $\lambda_f^+ > 0$ and $\lambda_f^- > 0$ is excluded from the computational process of Model (2). All the other (λ_f^+ and λ_f^-) combinations become feasible in both Equations (4) and (5), so being feasible in Model (2).

At the end of the first stage, it is necessary to separate all observations into multiple subsets. To explain the classification, let $\lambda_f^* (= \lambda_f^{+*} - \lambda_f^{-*})$, d^* and s^* be an optimal solution of Model (2). Then, the original data set (G) is classified into the following subsets $G = G_1 \cup G_2 = C_1 \cup D_1 \cup C_2 \cup D_2$ where

$$C_1 = \left\{ j \in G_1 \middle| \sum_{f=1}^{h} \lambda_f^* z_{fj} > d^* + s^* \right\},\ C_2 = \left\{ j \in G_2 \middle| \sum_{f=1}^{h} \lambda_f^* z_{fj} < d^* - s^* \right\},$$

$D_1 = G_1 - C_1$ and $D_2 = G_2 - C_2$.

Based upon the result, we determine that observations in C_1 belong to G_1 and those of C_2 belongs to G_2 because those observations locate clearly above or below an overlap identified from Model (2). The two subsets $(D_1 \cup D_2)$ consist of the overlap whose observations are not yet classified in the first stage.

Stage 2: In a case when two groups have an overlap, we need to reclassify all the observations in the overlap $(D_1 \cup D_2)$, because the group membership of these observations is still unknown and needs to be undetermined in the stage. Mathematically, the second stage has the following formulation:

$$\min \left\{ \sum_{j \in D_1} y_j + \sum_{j \in D_2} y_j \;\middle|\; \begin{array}{l} \sum_{f=1}^{h} (\lambda_f^+ - \lambda_f^-) z_{fj} - c + M y_j \ge 0,\ j \in D_1,\ \sum_{f=1}^{h} (\lambda_f^+ - \lambda_f^-) z_{fj} - c - M y_j \le -\varepsilon,\ j \in D_2, \\ \sum_{f=1}^{h} \left(\lambda_f^+ + \lambda_f^- \right) = 1,\ \zeta_f^+ \ge \lambda_f^+ \ge \varepsilon \zeta_f^+ \text{ (for all f)},\ \zeta_f^- \ge \lambda_f^- \ge \varepsilon \zeta_f^- \text{ (for all f)}, \\ \zeta_f^+ + \zeta_f^- \le 1 \text{ (for all f)},\ \zeta_f^+ + \zeta_f^- \ge \varepsilon \text{ (for some f)}, \\ c : \text{URS},\ \zeta_f^+, \zeta_f^- \ \&\ y_j : \text{binary, and all other varables} \ge 0. \end{array} \right\}, \quad (6)$$

where, the binary variable (y_j) counts the number of observations classified incorrectly. The objective function minimizes the number of misclassifications. We need to prescribe a large number (M) in Model (6). A discriminant score (c) is newly incorporated into Model (6) as an unknown unrestricted variable. The new score for the second stage substitutes for the previous discriminant score (d) in Model (2) for the first stage.

Non-Zero Condition (NZC): A possibility, to which we need to pay attention, is a simultaneous occurrence of $\lambda_f^+ = 0$ and $\lambda_f^- = 0$ in Model (6). The occurrence of zeros in the paired variables does not imply a mathematical problem as the computational result of Model (2). However, to control the number of positive λ_f estimates, we may add the following Non-Zero Condition (NZC) in (6):

$$NZC:\ \lambda_f^+ + \lambda_f^- \ge \varepsilon \text{ (for some f)}. \quad (7)$$

Here, several pairs avoid a simultaneous occurrence of $\lambda_f^+ = 0$ and $\lambda_f^- = 0$. Of course, we can specify all the pairs can avoid the condition of Equation (7). It depends upon the degree of freedom in the second stage of DEA-DA.

After obtaining an optimal solution on λ_f^* and c^* from Model (6), the second stage classifies observations in the overlap as follows: if $\sum_{f=1}^{h} \lambda_f^* z_{fj} \geq c^*$, then the j-th observation belongs to G_1 or if $\sum_{f=1}^{h} \lambda_f^* z_{fj} \leq c^* - \varepsilon$, then it belongs to G_2. Consequently, all the observations in G are classified into either G_1 or G_2.

CORPORATE VALUE MEASURE BY TOBIN'S Q

Tobin's *q* was initially part of an investment theory (Brainard and Tobin 1968; and Tobin 1969, 1978). In the late 1960's and 1970's, the research on investment was dominated by the neoclassical theory by Jorgenson (1963) and the theory of Tobin's *q*. The capital investment in the neoclassical theory implied a financial resource allocation for maximizing the corporate value of a firm. Thus, firms make their capital investments to obtain the optimal capital allocation for maximizing their profits.

Meanwhile, the optimal investment based upon Tobin's *q* theory is determined by a ratio that is referred to as "Tobin's *q*". The score indicates the ratio of the market value of a firm to the replacement cost of its physical assets. If Tobin's *q* is greater than unity, then the additional investment of a firm may create a new corporate value in a financial market. Conversely, if the ratio is less than unity, the new capital investment is not profitable.

Hayashi (1982) initiated an important theoretical work on Tobin's q. He has discussed that the two theories (i.e., the neoclassical theory of investment and the q theory suggested by Tobin) are mathematically equivalent under assumptions such as perfectly competitive markets and linearly homogeneous technology. In addition, his study has compared marginal *q* with average *q* and provided conditions under which average q is equal to marginal q. The marginal *q* is the ratio of a change in the corporate value of a firm to the added capital cost for a small increment to a capital stock. An average *q* is the same as the *q* ratio explained above. His research theoretically explains that marginal *q* should be a predictor for the new capital investment instead of average *q*. However, it is true that most of empirical studies have employed average *q* due to a difficulty on the estimation of marginal *q*. Thus, this research uses the average *q* as a measure of Tobin's *q*.

Since the 1980s, Tobin's *q* has explained a wide variety of corporate phenomena. Table 1 summarizes previous studies using Tobin's *q* as a measure of corporate performance. The row in Table 1 indicates the unique features of Tobin's q as a performance measure: (a) corporate opportunities (i.e., growth opportunities and investment opportunities), (b) economic return (i.e., from announcement effect, (c) differences in board composition, (d) corporate diversification, (e) dividend changes, (f) an existence of hedging strategy, (g) differences in ownership structure and (f) economic return from intangible assets (i.e., advertising stock or investment, brand equity, customer assets, economic rent, patent stock and R&D stock or investment). The column in Table 1 indicates types of calculation procedure of Tobin's q. All categorical procedures are primarily measured by Tobin's q = Market Value/ Replacement Value.

Table 1. Literature Review Regarding Tobin's s q

		Calculation Procedure of Tobin's q			
Feature of Tobin's q		Lindenberg and Ross	Hall	Simple	Lewellen and Badrinath
Corporate Opportunity	Growth Opportunities	Lang *et al*. (1991)		Wilbon (1999) Jiraporn (2006)	Gay and Nam (1998)
	Investment Opportunities	Lang *et al*. (1991)	Barnett and Sakellaris (1998)	Opler and Titman (1993) Schlingemann (2004)	Missonier-Piera (2007)
Economic return	Announcement Effects	Vafeas and Joy (1995) Szewczyk *et al*. (1996)		Min and Prather (2001) Brailsford and Yeoh (2004)	
	Board Composition	Yermack (1996)		Bhagat and Black (2001) Hossain *et al*. (2001) Prevost *et al*. (2002)	Bhagat and Black (2001)
	Corporate Diversification	Wernerfelt and Montgomery (1988) Lang and Stulz (1994) Berger and Ofek (1995)	Lloyd and Jahera (1994)	Chen and Ho (2000) Hyland (2003)	Ahn and Denis (2004)
	Dividend Changes	Ryan *et al*. (2000)	Lang and Litzenberger (1989) Denis *et al*. (1994)	Best and Best (2001) Gugler and Yurtoglu (2003)	
	Hedging				Nelson *et al*. (2005)
	Ownership Structure	McConnell and Servaes (1990) Cho (1998) Hermalin and Weisbach (1991)	Morck *et al*. (1988)	Park and Song (1995) Himmelberg *et al*. (1999) Demsetz and Villalonga (2001)	
	Takeover	Servaes (1991)	Servaes (1991)	Burns *et al*. (1998) Moeller *et al*. (2004)	
Economic return from intangible asset	Advertising Stock or			Bharadwaj *et al*. (1999) Chin *et al*. (2006)	
	Brand equity	Simon and Sullivan (1993)	Simon and Sulivan (1993)	Rao *et al*. (2004)	
	Customer assets			Anderson *et al*. (2004) Morgan and Rego (2006)	
	Economic rent	Lindenberg and Ross (1981) Salinger (1984) Smirlock *et al*. (1984)		Acquaah and Chi (2007)	
	Patent Stock			Hall *et al*. (2001) Chin *et al*. (2006)	
	R&D Stock or Investment	Jose *et al*. (1986)		Bharadwaj *et al*. (1999) Hsieh *et al*. (2003) Chin *et al*. (2006)	

The differences among categorical procedures are found in the measure of replacement value in the equation. The first row of Table 1 indicates the methodologies to compute Tobin's q. Generally, the computation of Tobin's q has four groups of approach: (a) the method proposed by Lindenberg and Ross (1981), (b) the method proposed by Hall *et al.* (1988), (c) the simple method proposed by Chung and Pruitt (1994) and Perfect and Wiles (1994), (d) the method proposed by Lewellen and Badrinath (1997).

This study examines the impact of R&D expenditure on the corporate performance measured by Tobin's q. The ratio is calculated by the simple calculation method (the sum of capitalized market value of total equity and debt, divided by book value of total assets). Thus, the position of this research is at the cell across "Simple" and "R&D Stock or Investment" in Table 1.

The literature review related to R&D field indicates the important role of R&D in the corporate performance. Jose *et al.* (1986) revealed that excessive R&D expenditure compared to the industry norm diminished the corporate performance. They employed multiple regression analysis to assess Tobin's q as a dependent variable. Bharadwaj *et al.* (1999) examined the relationship between Tobin's q and IT (information technology) related R&D expenditure by using hierarchical regression analysis.They concluded that there was positive relationship between them. Hsieh *et al.* (2003) investigated the impact of the R&D expenditure on the corporate performance in the pharmaceutical and chemical industries based on multiple regression analysis. They revealed that R&D investment might significantly improve the corporate performance. Chin *et al.* (2006) explored positive relationship between R&D spillover and Tobin's q in Taiwan semiconductor industry using multiple regression analysis.

As mentioned previously, this study uses Tobin's q as a corporate performance measure, based upon which we classify sample companies into two groups: companies with high corporate performance (high q) and those with low corporate performance (low q). Many researchers have employed Tobin's q as not only a measure of corporate performance but also a classification criterion (Oplar and Titman 1993, Min and Prather 2001). This study determines the criteria of high and low q in the following manner. If a firm is in the top 25% of Tobin's q, the firm belongs to the high q group. On the other hand, if the firm is in the bottom 25% of Tobin's q, the firm belongs to the low q group. This classification method can avoid a methodological drawback related to Tobin's q. It is widely known that the measurement of Tobin's q is fuzzy. The DEA-DA method along with the two group classification can avoid the fuzziness related to Tobin's q measurement by eliminating firms that this study has a difficulty in classifying them.

AN APPLICATION TO JAPANESE MANUFACTRUING INDUSTRY

Japanese manufacturing industries are often classified into "Industry of the 20th Century" and "Industry of the 21st Century". This study empirically compares between their influences of R&D expenditure on the financial performance of two Japanese manufacturing industries. In this study, the industry of the 20th century includes (a) nonferrous industry, (b) general machinery industry and (c) transportation machinery & apparatuses industry. Meanwhile, the industry of the 21st century includes (a) electrical machinery & apparatuses industry and (b) precision machinery & apparatuses industry. The research has the following two business assertions:

Business Assertion 1: The R&D expenditure enhances the corporate values of Japanese manufacturing firms.

Business Assertion 2: The influence of the R&D expenditure depends upon the type of Japanese manufacturing industry.

To investigate the above assertions, this study sampled the financial performances of 357 manufacturing firms from 2004 to 2007 (Japanese accounting period: April to March). The total sample size is 1071 (= 3 annual periods x 357 firms). Hereafter, this study refers to the industry of the 20th century as Type I and the industry of the 21st century as Type II. Type I industry includes 219 firms so that the total sample number of Type I is 657 (= 3 annual periods x 219 firms). Type II industry includes 138 firms so that the total sample number of Type II is 414 (= 3 annual periods x 138 firms).

The criterion to separate the two groups is Tobin's q, or so-called "simple q", which corresponds to [the market value of common equity – the book value of common equity + the total assets) / the total assets] at the end of each fiscal period. See Perfect and Wiles (1994) and Chung and Pruitt (1994). G_1 consists of firms with high q and G_2 consists of firms with low q.

Table 2. Descriptive statistics of Type I industry

	R&D I	NITR	LISE	Gror1	Exec	DPR
			High q Group			
Average	3.62	8.79	42.17	10.54	1.68	21.66
Median	2.98	7.02	24.76	9.18	0.16	18.20
Max.	13.46	99.97	279.27	105.14	19.91	177.38
Min.	0.00	-101.71	0.00	-95.70	0.00	-379.45
S.D.	2.87	20.04	50.23	22.98	3.65	38.51
			Low q Group			
Average	1.86	2.54	47.12	4.75	1.65	57.59
Median	1.38	3.29	16.84	5.16	0.64	17.90
Max.	7.25	46.15	505.47	53.48	18.91	5496.62
Min.	0.00	-67.78	0.00	-41.70	0.02	-105.71
S.D.	1.65	9.23	76.20	12.61	2.41	430.49
t test	6.78***	3.63***	-0.85	3.15***	0.09	-1.06

Note: The t-test at the bottom indicates the result of Welch's t test. The superscript *** denies a null hypothesis (there is no difference between two groups) at the level of 1% significance.

Source: Sueyoshi and Goto (2011).

This study uses six financial factors for the proposed financial analysis. First, R&D I (R&D Intensity): R&D expenditure divided by total revenue (%)[1]. Second, NITR (Ratio of net income to total revenue: net income divided by total revenue (%)[2]. Third, LISE (Ratio of

[1] McConnell and Servaes (1995) used the measure (R&D expenditure/total asset) to explain Tobin'q.

[2] Servaes (1996) used NITR to explain Tobin's q. See Yarmack (1997).

liability with interest to shareholder's equity: total liability with interest divided by shareholder's equity (%)[3]. Fourth, Grorl (Growth rate of total revenue on 1 year: (total revenue in this period – total revenue in the previous period) / total revenue in this period (%)[4]. Fifth, Exec (Ratio of executive shareholdings to total shares: executive shareholdings divided by total shares (%)[5]. Sixth, DPR (Dividend payout ratio: total dividend divided by revenue before tax (%)[6]. Tables 2 and 3 provide statistical data characterizations on Type I and Type II industries.

Table 3. Descriptive statistics of Type II industry

	R&D I	NITR	LISE	Grorl	Exec	DPR
			High q Group			
Average	7.02	13.89	125.63	31.40	4.53	9.42
Median	5.43	10.00	12.92	7.26	0.58	15.79
Max.	53.17	146.15	9627.63	2466.40	42.69	400.31
Min.	0.00	-44.94	0.00	-96.60	0.00	-1899.55
S.D.	7.54	21.37	942.28	242.64	8.43	200.25
			Low q Group			
Average	4.52	4.05	66.94	2.42	5.54	38.52
Median	4.09	3.88	17.37	1.85	0.82	23.62
Max.	19.37	20.49	4189.20	23.25	41.99	822.22
Min.	0.00	-29.20	-150.11	-63.23	0.03	-40.01
S.D.	3.29	6.24	409.81	9.96	10.09	91.59
t test	3.04***	4.53***	0.57	1.22	-0.78	-1.37

Note: The t-test at the bottom indicates the result of Welch's t test. The superscript *** denies a null hypothesis (there is no difference between two groups) at the level of 1% significance.

Source: Sueyoshi and Goto (2011).

Table 4 summarizes resulting parameter estimates of DEA-DA and logit regression. This study uses the logit regression as a methodological alternative of the proposed DEA-DA. This type of methodological comparision is important for any empirical study in avoiding a methodological bias, where it implies that different methods produce different result.

The bottom of Table 4 indicates the classification rate (i.e., the number of observations classified correctly divided by the number of total observations). As summarized in Table 4, DEA-DA outperforms logit regression in terms of the classification rates. For example, DEA-DA has 76.52% and 78.37% as the classification rates for Type I and Type II industries, respectively. Meanwhile, the logit regression has 69.82% and 69.71% for the two cases. This

[3] Jensen (1986, 1989) has discussed a disciplinary influence of debt that it may protect a firm from excess investment by corporate executives. Hence, this study uses LISE to explain Tobin's q.

[4] Groerl is due to the previous research of La Porta et al. (2002). See Connolly and Hirschey (2005), as well.

[5] Exec is due to Morck (1988) and Mcconell and Servaes (1995).

[6] DPR is due to the hypothesis of free cash flow proposed by Jensen (1986). Furthermore, an increase in dividend may enhance future profitability. This hypothesis was supported by Bhattacharya (1979), Miller and Rock (1985) and John and Williams (1985). In contrast, Watts (1973), Benartzi, Michaely and Thaler (1997), Grullon, Michaely and Swaminathan (2002), Grullon, Michaely, Benartzi and Thaler (2005) did not support the assertion. An exception was Nissim and Ziv (2001) that found a positive relationship between dividend and future profitability. See Denis, Denis and Sarin (1994), as well.

study does not deny the importance of the logit regression. Rather, this study conveys a message that an empirical study needs to consider the existence of a methodological bias.

This study finds the following results from Table 4: First, R&D I and NITR have same signs between DEA-DA and logit regression in the two industries. Exec and DPR have same signs between the two methods in Type I industry. LISE and Grorl have same signs between the two methods in Type II industry. Second, DEA-DA in the second stage has high weights on R&D I, NITR, LISE and Exec in Type I industry. The logit regression identifies a statistical significance on R&D I and NITR in the two industries. Third, the second stage of DEA-DA has high weights on R&D I, NITR, Grorl and DPR in Type II industry. The logit regression identifies a statistical significance on R&D I and NIOR in Type II industry.

Table 4. Results of DEA-DA and Logit estimation

Industries		Type I Industry			Type II Industry		
Models		DEA-DA 1 Stage	DEA-DA 2 Stage	Logit	DEA-DA 1 Stage	DEA-DA 2 Stage	Logit
		Weights / Parameters					
Variables	Discriminant Score / Constant	-0.035	-0.056	-0.028	-0.185	-0.013	0.262
	R&D I	0.107	0.207	1.070*** (0.268)	0.091	0.102	0.932*** (0.229)
	NITR	0.022	0.408	0.690*** (0.172)	0.161	0.256	1.338*** (0.329)
	LISE	-0.129	0.205	0.186 (0.046)	0.038	0.000	0.245 (0.060)
	Grorl	-0.036	0.027	0.378*** (0.094)	0.555	0.347	1.571 (0.386)
	Exec	-0.038	-0.154	-0.108 (-0.027)	-0.070	0.000	-0.256 (-0.063)
	DPR	-0.667	-0.000	-1.676* (-0.419)	-0.086	0.295	-0.268 (-0.066)
Classification Rate	High q Firms	81.10%		65.24%	69.23%		66.35%
	Low q Firms	71.95%		74.39%	87.50%		73.08%
	Total Firms	76.52%		69.82%	78.37%		69.71%

Note: the number within ( ) indicates a marginal effect of each parameter measured by logit regression. For example, the marginal effect (0.229) of R&D I indicates that if one unit increase in R&D Intensity increase the probability that an observation is classified into G1 (group with high q). Source: Sueyoshi and Goto (2011).

Business Implications: The empirical findings have the business implication that NITR and R&D I have a positive influence on the corporate value (measured by Tobin's q) of Type I (Industry of the 20th Century) and Type II (Industry of the 21st Century). Thus, the R&D expenditure is important for the two Japanese manufacturing industries. So, this study confirms the first business assertion. Moreover, the importance of R&D expenditure is different between the two types of industries. In particular, the importance in Type I industry is larger than that of Type II industry. Thus, this study confirms the validity of the second business assertion. In addition, it is important to note that the Type II industry needs high level of growth and dividend to enhance its corporate value.

CONCLUSION AND FUTURE EXTENSION

This study investigated the influence of R&D expenditure on the corporate value of Japanese manufacturing industry. DEA-DA and logit regression were methodologies used for this study. This research found that the R&D expenditure was important for the Japanese manufacturing industries to be competitive in a global market. However, the importance of R&D expenditure in Type I industry was larger than that of Type II industry because Type I was a technologically matured industry and Type II was an industry in which technology innovations have been constantly occurred. As a result, the R&D expenditure was more influential on the Type I industry than the Type II industry. In this sense, the Type II was a high risk and high return industry.

Japanese manufacturing firms have been long believed that they can be a major play in a global market as long as they focus upon technology innovation. As an important corporate strategy, they have invested a large amount of R&D expenditure. Such a view needs to be changed in a global market. The message to the audience is that technology innovation is indeed important. However, a marketing effort in short and long terms is as important as the technology innovation in order to survive in the global market.

Finally, it is hoped that this study contributes in technology management and we look forward to seeing future extensions as discuused in this study.

REFERENCES

Acquaah, M., Chi, T. 2007. A longitudinal analysis of the impact of firm resources and industry characteristics on firm-specific profitability, *Journal of Management and Governance*, 11: 179-213.

Ahn, S., Denis, D.J. 2004. Internal capital markets and investment policy: evidence from corporate spinoffs, *Journal of Financial Economics*, 71: 489-516.

Anderson, E.W., Fornell, C., Mazvancheryl, S.K.. 2004. Customer Satisfaction and Shareholder Value, *Journal of Marketing*, 68: 172-185.

Bajgier, S.M., Hill, A.H. 1982. A comparison of statistical and linear programming approaches to the discriminant problem, *Decision Sciences*, 13: 604-618.

Barnett, S.A., Sakellaris, P. 1998. Non-linear response of firm investment to Q: testing a model of convex and non-convex adjustment, *Journal of Monetary Economics*, 42: 261-288.

Benartzi, S., Michaely, R., Thaler, R.H. 1997. Do changes in dividends signal the future or the past?, *Journal of Finance*, 52: 1007-1034.

Berger, P.G.' Ofek, E. 1995. Diversification's effect on firm value, *Journal of Financial Economics,* 37: 39-65.

Best, R.J., Best, R.W. 2001. Prior Information and the Market Reaction to Dividend Changes, *Review of Quantitative Finance and Accounting*, 17: 361-376.

Bhagat, S., Black, B. 2001. The non-correlation between board independence and long term firm performance, *Journal of Corporation*, 27: 231–274.

Bharadwaj, A.S., Bharadwaj, S.G., Konsynski, B.R. 1999. Information technology effects on firm performance as measured by Tobin's q, *Management Science,* 45: 1008–1024.

Bhattacharya, S. 1979. Imperfect information, dividend policy, and "the bird in the hand" fallacy, *Bell Journal of Economics*, 10: 259-270.

Brailsford, T.J., Yeoh, D. 2004. Agency problems and capital expenditure annoucements, *The Journal of Business,* 77: 223-256.

Brainard, W.C., Tobin, J. 1968. Pitfalls in financial model building, *The American Economic Review*, 58: 99-122

Burns, R., Erwin, G., Messina, F., Nail, L. 1998. Value creation in bundling utility mergers: a corporate focus anomaly, *Journal of Energy Finance & Development*, 3: 185-192.

Charnes, A., Cooper, W.W. 1977. Goal programming and multiple objective optimization, *European Journal of Operational Research,* 1: 39-54.

Charnes, A., Cooper, W.W., Ferguson, R.O. 1955. Optimal estimation of executive compensation by linear programming, *Management Science*, 1: 138-151.

Chen, S., Ho, K.W. 2000. Corporate diversification, ownership structure, and firm value: The Singapore evidence, *International Review of Financial Analysis*, 9: 315–326.

Chin, C.-L., Lee, P., Chi, H.-Y., Anandarajan, A. 2006. Patent citation, R&D spillover, and Tobin's Q: Evidence from Taiwan semiconductor industry, *Review of Quantitative Finance and Accounting*, 26 : 67-84

Cho, M.-H. 1998. Ownership structure, investment, and the corporate value: An empirical analysis, *Journal of Financial Economics*, 47: 103-121.

Chung, K.H., Pruitt, S.W. 1994. A simple approximation of Tobin's q, *Financial Management*, 23: 70-74.

Connolly, R.A., Hirschey, M. 2005. Firm size and the effect of R&D on Tobin's q, *R&D Management*, 35: 217-223.

Demsetz, H., Villalonga, B. 2001. Ownership structure and corporate performance, *Journal of Corporate Finance*, 7: 209-233.

Denis, D., Denis, D., Sarin, A. 1994. The information content of dividend changes: Cash flow signaling, overinvestment, and dividend clienteles, *Journal of Financial and Quantitative Analysis,* 29: 567– 587.

Doumpos, M., Zanakis, S.H., Zopounidis, C. 2001. Multicriteria preference disaggregation for classification problems with an application to global investing risk, *Decision Sciences*, 32: 333-385.

Duarte Silva, A.P., Stam, A. 1994. Second-order mathematical programming formulations for discriminant analysis, *European Journal of Operational Research*, 72: 4-22.

Falk, J.E., Karlov, V.K. 2001. Robust separation of finite sets via quadratics, *Computers and Operations Research*, 28: 537-561.

Fisher, R.A. 1936. The use of multiple measurements in taxonomy problems, *Annals of Eugenics*, 7: 179-188.

Freed, N., Glover, F. 1981a. Simple but powerful goal programming models for discriminant problems, *European Journal of Operational Research*, 7: 44-60.

Freed, N., Glover, F. 1981b. A linear programming approach to the discriminant problem, *Decision Sciences*, 12: 68-74.

Gay, G.D., Nam, J. 1998. The underinvestment problem and corporate derivatives use, *Financial Management*, 27: 53-69.

Grullon, G., Michaely, R., Benartzi, S., Thaler, R.H. 2005. Dividend changes do not signal changes in future profitability, *Journal of Business*, 78: 1659-1682.

Grullon, G., Michaely, R., Swaminathan, B. 2002. Are dividend changes a sign of firm maturity?, *Journal of Business*, 75: 387-424.

Gugler, K., Yurtoglu, B.B. 2003. Corporate governance and dividend pay-out policy in Germany, *European Economic Review*, 47: 731-758.

Hall, B., Cummins, C., Laderman, E., Mundy, J. 1988. *The R&D master file documentation, Technical working paper No. 72,* National Bureau of Economic Research.

Hall, B., Jaffe, A., Trajtenberg, M. 2001. *Market Value and Patent Citations: A First Look, Working paper*, University of California, Berkeley.

Hayashi, F. 1982. Tobin's q and average q: A neoclassical interpretation, *Econometrica*, 50: 213-224

Hermalin, B., Weisbach, M. 1991. The effects of board composition and direct incentives on firm performance, *Financial Management*, 20: 101-12.

Himmelberg, C.P., Hubbard, R.G., Palia, D. 1999. Understanding the determinants of managerial ownership and the link between ownership and performance, *Journal of Financial Economics*, 53: 353-384.

Hossain, M., Prevost, A.K., Rao, R.P. 2001. Corporate governance in New Zealand: The effect of the 1993 Companies Act on the relation between board composition and firm performance, *Pacific-Basin Finance Journal*, 9: 119–145.

Hsieh, P.-H., Mishra, C.S., Gobeli, D.H. 2003. The return on R&D versus capital expenditures in pharmaceutical and chemical industries, *IEEE Transactions on Engineering Management*, 50: 141-150.

Hyland, D.C. 2003 The effect of diversification of firm value: A pre-and post-diversification analysis, *Studies in Economics and Finance*, 21: 22-39.

Jensen, M.C. 1986. Agency costs of free cash flow, corporate finance and takeovers, *American Economic Review*, 76: 323-329.

Jensen, M.C. 1989. Eclipse of the public corporation, *Harvard Business Review*, 67: 61-74.

Jiraporn, P. 2006. Share repurchases, shareholder rights, and corporate governance provisions, *North American Journal of Economics and Finance*, 17: 35–47.

John, K., Williams, J. 1985. Dividends, dilution, and taxes: a signaling equilibrium, *Journal of Finance,* 40: 1053-1070.

Jorgenson, D.W. 1963 Capital theory and investment behavior, *The American Economic Review*, 53: 247-59

Jose, M.L., Nichols, L.M., Stevens, J.L. 1986. Contributions of diversification, promotion, and R&D to the value of multiproduct firms: A Tobin's q approach, *Financial Management*, 15: 33-42.

Lang, L., Litzenberger, R.H., 1989, Dividend announcements cash flow signalling vs free cash flow hypothesis, *Journal of Financial Economics*, 24: 181-191.

Lang, L., Stulz, R.M. 1994. Tobin's q, Corporate Diversification, and Firm Performance, *The Journal of Political Economy*, 102: 1248-1280.

Lang, L., Stulz, R.M., Walking, R. 1991. A test of the free cash flow hypothesis: The case of bidder returns, *Journal of Financial Economics*, 29: 315–336.

La Porta, R., Lopez-de-Silanes, F., Shleifer, A., Vishny, R. 2002. Investor protection and corporate valuation, *Journal of Finance*, 57: 1147-1170.

Lewellen, W.G., Badrinath, S.G. 1997, On the measurement of Tobin's q, *Journal of Financial Economics*, 44: 77-122.

Lindenberg, E.B., Ross, S.A. 1981. Tobin's q ratio and industrial organization, *Journal of Business*, 54: 1-32.

Lloyd, W.P., Jahera, Jr. J.S. 1994. Firm-diversification effects on performance as measured by Tobin's q, *Managerial and Decision Economics*, 15: 259-266.

Maddala, G. S. 1983. *Limited Dependent Variables and Qualitative Variables in Econometrics*, Cambridge Univ. Press, New York.

Mangasarian, O.L. 1999. Arbitrary-norm separating plane, *Operations Research Letters*, 24: 15-23.

McConnell, J.J., Servaes, H. 1990. Additional evidence on equity ownership and corporate value, *Journal of Financial Economics*, 27: 595-612.

McConnell, J.J., Servaes, H. 1995. Equity ownership and the two faces of debt, *Journal of Financial Economics*, 39: 131-157.

McFadden, D. 1980. Econometric models for probabilistic choice, *Journal of Business*, 53: 513-529.

McLachlan, G.L. 1992. *Discriminant Analysis and Statistical Pattern Recognition*, John Wiley, New York.

Miller, M.H., Rock, K. 1985. Dividend policy under asymmetric information *Journal of Finance*, 40: 1031-1051

Min, J.H., Prather, L.J. 2001 Tobin's q, agency conflicts, and differential wealth effects of international joint ventures, *Global Finance Journal*, 12: 267–283.

Missonier-Piera, F. 2007. Motives for fixed-asset revaluation: An empirical analysis with Swiss data, *The International Journal of Accounting*, 42: 186-205.

Moeller, S.B., Schlingemann, F.P., Stulz, R.M. 2004. Firm size and the gains from acquisitions, *Journal of Financial Economics*, 73: 201-228.

Morck, R., Shleifer, A., Vishny, R.W. 1988. Management ownership and market valuation", *Journal of Financial Economics*, 20: 293-315.

Morgan, Neil A., Rego, L.L. 2006. The value of different customer and loyalty metrics in predicting business performance, *Marketing Science*, 25: 426–39.

Nelson, J.M., Moffitt, J.S., Affleck-Graves, J. 2005. The impact of hedging on the market value of equity, *Journal of Corporate Finance*, 11: 851-881.

Nissim, D., Ziv, A. 2001. Dividend changes and future profitability, *Journal of Finance*, 56: 2111-2133.

Opler, T., Titman, S. 1993. The determinants of leveraged buyout activity: free cash flow vs. financial distress costs, *The Journal of Finance*, 48: 1985-1999.

Park, S., Song, M.H. 1995. Employee stock ownership plans, firm performance, and monitoring by outside blockholders, *Financial Management*, 24: 52-65.

Perfect, S.B., Wiles, K.W. 1994. Alternative constructions of Tobin's q: An empirical comparison, *Journal of Empirical Finance*, 1: 313-341.

Prevost, A.K., Rao, R.P., Hossain, M. 2002. Determinants of board composition in New Zealand: a simultaneous equation approach, *Journal of Empirical Finance*, 9: 373–397.

Rao, V.R., Agarwal, M.K., Dahlhoff, D. 2004. How is manifest branding strategy related to the intangible value of a corporation?, *Journal of marketing*, 68: 126-141.

Ryan, P.A., Besley, S., Lee, H.W. 2000. An empirical analysis of reactions to dividend policy changes for NASDAQ firms, *Journal of Financial and Strategic Decisions*, 13: 35-44.

Salinger, M.A. 1984. Tobin's q, unionization, and the concentration-profits relationship, *Rand Journal of Economics*, 15: 159-170.

Schlingemann, F.P. 2004. Financing decisions and bidder gains, *Journal of Corporate Finance*, 10: 683-701.

Servaes, H. 1991. Tobin's Q and the gains from takeover, *The Journal of Finance*, 46: 409-419.

Servaes, H. 1996. The value of diversification during the conglomerate merger wave, *Journal of Finance*, 51: 1201-1225.

Simon, C.J., Sullivan, M.W. 1993. The measurement and determinants of brand equity: a financial approach, *Marketing Science*, 12: 28-52.

Smirlock, M., Gilligan, T., Marshall, W. 1984. Tobin's q and the structure-performance relationship, *The American Economic Review*, 74: 1051-1060.

Smith, C.A.B. 1947. Some examples of discrimination, *Annals of Eugenics*, 13: 272-282.

Sueyoshi, T. 2004. Mixed integer programming approach of extended-discriminant analysis, *European Journal of Operational Research*, 152: 45-55.

Sueyoshi, T. 2005. Financial ratio analysis of the electric power industry, *Asia-Pacific Journal of Operational Research*, 22: 349-376.

Sueyoshi, T. 2006. DEA-discriminant analysis: methodological comparison among eight discriminant analysis approaches, *European Journal of Operational Research*, 169: 247-272.

Sueyoshi, T. Goto, M. 2011. DEA-DA (discriminant analysis) to measure importance of R&D expenditure in information technology (IT) industry: comparative analysis between Japanese IT and manufacturing industries. *Working Paper 2011-007*. New Mexico Insitute of Mining and Technology, Department of Management, 801 Lery Place, Soccro, NM.

Szewczyk, S.H., Tsetsekos, G.P., Zantout, Z. 1996. The valuation of corporate R&D expenditures: evidence from investment opportunities and free cash flow, *Financial Management*, 25: 105-110.

Tobin, J. 1969. A general equilibrium approach to monetary theory, *Journal of Money, Credit and Banking*, 1: 15-29.

Tobin, J. 1978. Monetary policies and the economy: the transmission mechanism, *Southern Economic Journal*, 44: 421-431.

Vafeas, N., Joy, O.M. 1995. Open market share repurchases and the free cash flow hypothesis G35, *Economics Letters*, 48: 405-410.

Watts, R. 1973. The information content of dividends, *Journal of Business*, 46: 191-211.

Wernerfelt, B.,Montgomery, C.A. 1988. Tobin's q and the importance of focus in firm performance, *The American Economic Review*, 78: 246-250.

Wilbon, A.D. 1999. An empirical investigation of technology strategy in computer software initial public offering firms, *Journal of Engineering and Technology Management*, 16: 147-169.

Yanev, N., Balev, S. 1999. A combinatorial approach to the classification problem, *European Journal of Operational Research*, 115: 339-350.

Yermack, D. 1996. Higher market valuation of companies with a small board of directors, *Journal of Financial Economics*, 40: 185-211.

Zopounidis, C., Doumpos, M. 2002. Multicriteria classification and sorting methods: a literature review, *European Journal of Operational Research*, 138: 229-246.

In: Economics of Innovation, Incentives and Uncertainty ISBN: 978-1-62257-251-9
Editors: P. Simmons and S. T. Jordan

Chapter 5

SYSTEM OF TRADE PREFERENCES AND INDUSTRIAL DEVELOPMENT IN BENEFICIARY ECONOMY: A VINTAGE FRAMEWORK FOR IMPACT STUDY

Abdelrasaq Nal
Tshwane University of Technology, Pretoria, South Africa

ABSTRACT

This chapter argues that current approaches to studying the impact of trade system preferences (non-reciprocal) on development outcomes in beneficiary economies have not been quite comprehensive in coverage. Especially with respect to the objective of promoting industrial development, most studies rely only on static impact issue of export facilitation to pass judgment on success of program implementation. The dynamic aspect of the analysis has largely been ignored. Combining insights from the literature on trade preferences and globalization, the present chapter proposes a three-dimensional impact analysis model that incorporates elements of both static and dynamic issues. These issues concern the roles the program can play in, facilitating exports, enhancing competitiveness of producers and fostering emergence of linkage related integrated manufacturing. It is hoped that a vintage framework of this sort will help guide future endeavors towards more comprehensive analytical engagements.

1. INTRODUCTION

Generally speaking, the perception that industrialization is the fulcrum of rapid growth and development in capitalist societies is hardly assailable in its macroeconomic fundamentals. Also, in terms of lessons from history, evidence points in the direction that structural transformation from traditional agriculture to modern industry is the key to fostering economic development. Whether it is the story of eighteenth century Britain's growth performance that is the issue, or the catch-up rhetoric of Korea and Japan in the twentieth century, or further still, the recent 'growth miracle' of 'Asian Tigers' that interests us, a constant theme in all is that industrialization played a central role in recorded achievements. Thus, it is hardly likely to surprise even the most casual observers that strategic development intervention programs in developing economies usually have economic diversification through creation of a virile industrial sector as a central goal.

A good example of this is the system of trade preferences for developing countries (non-reciprocal trade preference programs) that has become a major feature of interaction between the industrialized North and the developing South since the late 1960s. As a trade related measure designed to support the industrialization efforts of the South, it provides special market access privileges for beneficiaries' exports to markets in industrialized donors' countries. Since these programs became operational nearly half a century ago, several attempts have been made to examine their impact on economic performances of beneficiaries. A review of this impact assessment literature however, reveals that the frameworks that have so far guided much of the investigations are not sufficiently revealing. Especially in relation to the goal of industrial development, much of these works can be criticized for restricting their focus only to the static issue of export facilitation. The potential dynamic returns in terms of efficiency upgrading and linkage development required for a robust development of the industrial sector have so far been given very scant attention. This is rather unfortunate especially given that the on-going restructuring of industrial production activities occasioned by the new wave of globalization is increasingly putting the dynamic issues at the forefront of the sustainable development agenda.

The objective in this chapter is to highlight the key issues, in the program impact assessment exercise, that are germane to the prospect of successful industrialization. This we achieved by developing and proposing a vintage framework for an all-inclusive analysis of the impact of the system of trade preferences on industrial development outcomes in beneficiary countries.

In the section that follows, the placement place of the system of trade preferences within the regulatory framework of the World Trade Organization (previously General Agreement on Tariffs and Trade (GATT)) is first discussed. This is followed by a treatment of the conceptual issues in Section 3. Approaches to empirical assessments and their shortcomings are then analyzed in Section 4 before the role of the new globalization in industrialization process is considered in Section 5. Finally in Section 6 the chapter's concluding remark is provided.

2. System of Trade Preferences within the Rules of the World Trade Organization

The principle of non-discrimination represents one of the fundamental ethical standards enshrined in the rules of the World Trade Organization (WTO) to govern its affairs. In its simplest form, it requires all contracting members of the organization to bind themselves to treating each other equally in the conduct of international trade. However, the realization that all trading partners are not equal and that a fair and equitable arrangement ought to provide special opportunities for the weak partners must have informed the decision of its authors to introduce some exceptions to this general rule. Under the current arrangement, there are three alternative provisions for trade discrimination within the GATT/WTO rules (Panagariya, 2000).

One, Article XX1V of the General Agreement allows two or more members of the WTO to form a free trade area (FTA) or a customs union (CU). This is the only provision where preferential arrangements which benefit developed countries such as European Economic Community (EEC) and North American Free Trade Area (NAFTA) are accommodated.

Two, under the enabling clause, developing countries can engage in any form of trade preferences among themselves. Because the objective is to promote trade among these countries, the preferences to which they agree need not be up to a Free Trade Area or Customs Union status. Preferential trade arrangements among developing countries such as Southern Africa Customs Union (SACU), Economic Community of West African States (ECOWAS), Southern Cone Common Market (MERCOSUR) and the ASEAN Free Trade Area (AFTA) belong to this category.

Three, Part IV of the GATT, added in 1964, contains measures designed to promote export from developing to developed countries. For example, Article XXXVI:8 provides that developed countries contracting parties do not expect reciprocity for commitments made by them in trade negotiations to reduce or remove tariffs and other barriers to trade of less-developed contracting parties. What this means is that developed countries can grant (on their own volition) one-way tariff concessions to goods imported from poor and underdeveloped countries. It is one-way in the sense that developing countries may not be required to reciprocate the gesture by reducing or removing their own barriers to imports of the same or similar goods from developed countries. It is within this framework that the systems of trade preferences for development are accommodated. European Union's Everything But Arms (EBA) Initiative and United States' African Growth Opportunity Act (AGOA) are prominent recent examples of the type of arrangements implied here.

3. Conceptual Issues

3.1. Motivations and Objectives

As earlier suggested, the key inspiration for the introduction of non-reciprocal systems of trade preferences is the acceptance that countries are generally unequally endowed in terms of economic power. While some are at advanced stage of development some are followers attempting to catch up, while a good number of others falls in the category of poor and underdeveloped nations. In order to assist the process of development in the less developed ones, the system of trade preferences for development programs was conceived and put in place in their favor. Basically, the program works by having a donor government (developed country) extend a one-way (non-reciprocal) preferential market access treatment, through reduction or elimination of artificial trade barriers, to her imports of goods originating from a beneficiary nation (developing country). The main objectives of this type of arrangement are:

a) to increase export earnings of beneficiaries;
b) to promote industrialization in beneficiary countries; and

c) to accelerate the rates of economic growth in beneficiary countries[1]

With respect to the objective of promoting industrialization, it is certainly very persuasive to argue that exports of developing countries' origin be given some non-reciprocal preferential treatments in bilateral trade between developed and developing countries. A major attribute of a typical developing economy is that its domestic market, relative to developed economies, is limited in size. As such production costs are on the high side for local firms as they are unable to reap the kind of efficiency returns that come with large-scale operations.

On the other hand, firms in developed countries with relatively larger internal markets are able to operate at higher levels of production efficiency than their counterparts. Any trade arrangement between these groups on a level playing field will undoubtedly be unfair as exports of firms from the developing countries will face serious obstacles penetrating markets of the developed countries. The same statement can however, not be made in reverse. On the contrary it is quite likely that exports of developed countries' origin would not only have little or no problem penetrating their counterparts' markets they may even go further to displace their products from the market. But if some measures of support, like waiving tariff payments for developing countries' exporters without necessarily obliging them to reciprocate can be put in place, it would be possible to restore some measures of balance to an otherwise uneven and one-sided contest.

3.2. Economic Mechanism

But in order to appreciate how the scheme can foster the development of a virile industrial sector in a beneficiary economy a demonstration of its economic mechanism will be needed.

Let us start by assuming a given developed country (say Country A) decides to extend preferential treatment to exports from a given developing country (say Country B). Key assumptions are that because of her lower level of development, the domestic market in country B is small while for the opposite reason country A has relatively large market. Due to the small size of country B's local market production costs of firms are positively off their international cut-off mark and as such their products could not compete with the developed country's products.

If country A now grants producers in country B special market access concessions by removing tariffs due on her imports from country B, an artificial cost advantage has been created for otherwise cost disadvantaged producers. As long as these concessions are significant enough to make penetrations into the larger market of preference granting donor preferences possible, it is expected that exports of items receiving preferences will increase. Thus our first proposition can be stated as:

> Export facilitation is the first dimension of potential effects of the program on industrial development and represents a static impact issue.

---

[1] This is as contained in the United Nations Conference on Trade and Development (UNCTAD)'s Declaration on the Generalized System of Preferences (GSP) for developing countries in New Delhi in 1968. See, Onguglo (1999) and UNCTAD (1985).

With flourishing export activities in country B, the product sector in question now becomes attractive to investors. Hence, money would flow into both existing and new plants sufficiently enough to make transformation from small scale to large scale operations possible. What comes with increased scale operations is scale economies that if successfully internalized will bring down the cost of production for beneficiary producers and enhance their international competitiveness. The second proposition is given as:

> Enhanced firm competitiveness through scale driven reduction in average cost of production is the second dimension of potential effects of the program on industrial development and represents an aspect of the dynamic impact issue.

In the final analysis, enhanced competitiveness on the part of manufacturing firms is believed to be what is needed for producers to out-grow protection afforded by preferential trade incentives. The whole process bears striking semblance to the famous inward oriented import substitution infant industry argument and can rightly qualify for description as its export/outward oriented version.

In Figure 1, an illustration of the mechanism just described is presented. Previously equilibrium is at point A with average cost *avc1* and industry size/scale *sc1*. With market access privileges and subsequent export surge (static gains) industry size/scale increases from *sc1* to *sc2* and this causes the average production cost to fall from *avc1* to *avc2* (dynamic returns). In consequence, a new equilibrium is established at point B with average cost *avc2* and industry size/scale *sc2*.

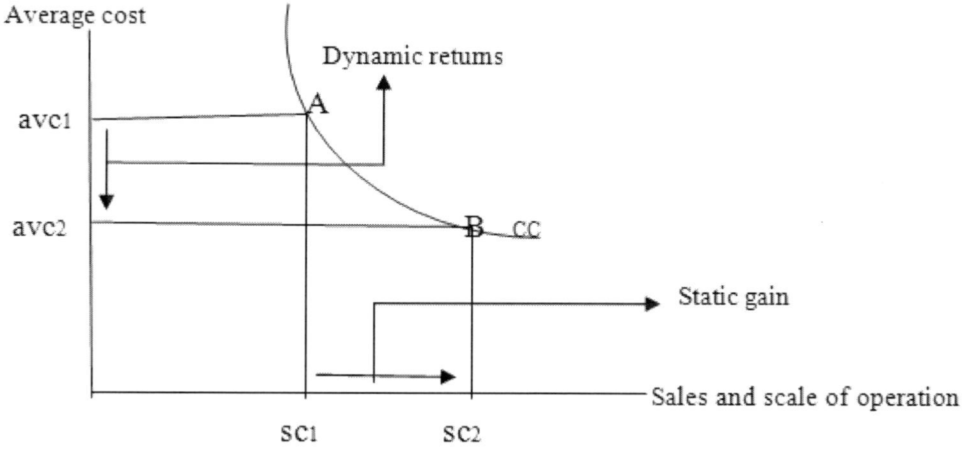

Figure 1. Effect of foreign market penetration brought about by preferential trade incentives.

Foreign market penetration, created by preferential trade incentives, leads to a simultaneous increase in sales and industry production scale. This is represented by a rightward shift on the horizontal axis from SC1 to SC2 (static gain). The result is a fall in average production cost as given by the downward shift on the vertical axis from avc1 to avc2 (dynamic return).

4. Empirical Approaches and their Shortcomings

The above illustration suggests that any analysis of the impact of system of trade preferences on industrial development outcome will involve considerations for both static and dynamic ends of gains. However, much of the works carried out so far in this area have been confined only to the former with little attention given to the latter.

Using ex ante data, one group of studies deployed the instrument of partial and general equilibrium analysis to simulate possible effects of programs on export performances of beneficiaries. A fairly recent example of these works can be found in Mattoo, et al, (2002) whose efforts were directed at assessing the quantitative impacts of the US AGOA scheme on African exports. Like all general and partial equilibrium studies that came before it, findings did not get beyond predicting the magnitude of these impacts under assumed scenarios.

Using ex post rather than ex ante data another group of research examines the impact of the programs by employing proxy for the trade incentive. The most popular instrument of analysis among this group of studies is the gravity model of trade. Typically, bilateral trade will be assumed to be a function of economic size and distance between trading partners. Then the model will be augmented by a suitable proxy for the incentive scheme to determine whether or not it is significant in explaining the observed trade flows. ,Usually, significant test statistics are commonly taken as proof of the positive impact of the program on the beneficiary's exports. Thus it is not uncommon to find, from works done by such gravity modelers like Sapir and Ldunberg (1984) on US GSP scheme, Nilsson (2002) on EU preferences to developing countries, Ledreman and Ozden (2004) on all US preferential programs, amongst others, results suggesting that the program had positive or no impact on export performances.

However, we find from Seyoum (2006) a concern for dynamic impact issue. But his attempt to accommodate this in work carried out using factor and multivariate regression analysis did not go beyond data treatment: more recent data which the author believed were capable of capturing the long term dimension of export supply responses to program incentives were employed. But like in other studies, this tendency to focus on export flow as yardstick for measuring the success or otherwise of preferential incentives grossly underestimates the complex nature of industrial development. It fails to reveal the underlying transformation in terms of productivity improvement required for performance to be sustainable independently of the trade incentive program in the long run. This message was delivered more poignantly in Rolfe and Woodward (2005). Based on their research findings that there were poor local contents in the AGOA facilitated garment exports from sub-Saharan Africa to the US, the authors concluded that the so-called export successes were driven less by competitive advantage than by trade protection that will erode with time. They gave the following remark:

> '... export value and growth, often used to gauge the successes of preferential trade agreements like AGOA, can be misleading.'

5. Contemporary Globalization and Prospects for Industrial Development in Developing Regions

The two dimensional framework to impact assessment suggested by the above mechanism can be stretched further if we consider how the new form of globalization is affecting prospects for industrial development in territorialized locations. This is taken up in details in sub-Sections 5.1, 5.2 and 5.3 below.

5.1. The New Globalization

It might be useful to start by asking what can be considered an obvious question: what is it that is new in the 'new or contemporary globalization'? While admitting that the phenomenon itself does not represent anything new, there is something novel however, in its contemporary form. The twentieth century had particularly served as the launch pad to some of the most important revolutions that the world ever witnessed in the areas of information, communication and production technologies. These developments were further complemented by significant liberalization of trade and investment regulations to bring about some dramatic transformations in the way individuals, businesses, and nations, interact with each other. The new form of interaction that emerged, in the view of Modeski, is characterized by widening and deepening of the systemic interdependencies amongst nations, civilizations and political communities (Modeski, 2005). This is what Dickens termed 'globalization' and he moved further to argue that it differs from its earlier episode which he referred to as 'internationalization' in a significant way: whereas internationalization involves no more than the simple extension of economic activities across national boundaries, reflecting essentially quantitative changes, globalization on the other hand, moves beyond mere geographical extension of economic activities to functional integration of such activities, reflecting mainly qualitative changes (Dickens, 2003).

In his own attempt to distill the nature of contemporary globalization from that which preceded it, Gereffi remarked that global interconnectedness in the past was characterized by shallow integration of activities manifested mainly in the form of trade in goods and services between independent firms and through international movements of portfolio capital. Its current form is, however, qualitatively different in that deep integration involving cross-border production of goods and services is its hallmark (Gereffi, 2005). The intensity of the process is such that production activities around goods are now assuming the dimension of global value-adding networks and flows (Held, et al., 2005).

Perhaps, a reflection on Dickens's illuminating description would be needed to help drive home the point better:

> "…Today, fewer and fewer industries are oriented towards local, regional or even national markets. A growing number of economic activities have meaning only in a global context. Thus, whereas a hundred or more years ago only rare and exotic products and some basic raw materials were involved in truly international trade, today virtually everything one can think of is involved in long-distance movement. And because of the increasingly complex ways in which production is organized across national boundaries, rather than contained

within them, the actual origin of individual products may be very difficult to ascertain." (Dickens, 2005).

5.2. Global Production Networking and Enclave Tendencies of Suppliers

The profound nature of the transformation described above by Dickens is such that it has inspired the emergence of some new paradigm in development literatures. In various attempts to give lucid interpretations to what some analysts observed to be going on, new concepts describing different bodies of knowledge, such as Value Chain, Global Commodity/Value Chain (GVC), Global Production Network (GPN) etc. have been added to the development lexicon.

A central insight of these new paradigms is the idea that the entire range of activities needed to produce goods of a certain kind is now being configured in a system of global inter-firm networks. Typically, a value activity, say product design, could take place in Country A, while Country B would take charge of the manufacturing end, before Country C performs the packaging or assemblage services and so on an so forth. Simply summed up, the whole arrangement reflects a system of globally dispersed but functionally integrated production networking whereby outputs from one country feeds in, as intermediate input, to the production process in another country.

To qualify for membership of a given network, fulfillments of certain eligibility criteria by aspiring nations are required. These are often determined, among others, by such national economic attributes as market potentials and relative efficiency of carrying out production activities in the region. Regions with a large internal market or good access to one, such as those conferred by trade preference schemes, are usually the preferred locations for the performance of much of the downstream end of production processes. For example, with respect to the latter, the shores of many African countries benefit from preferential trade development assistance programs because they are usually the preferred locations for carrying out assemblage functions like Cut Make & Trim (CMT) operations in the apparel value chain. Material and other resource endowments could also qualify one for membership insofar as these endowments confer, on producers, the competitive edge needed to be atop the productivity frontier of a value item relevant to operations of a network.

However, it is one thing to qualify for membership; it is another thing to remain relevant and sustain the membership. One of the defining attributes of the system of global networking from form of production is that the relationship is fluid: it is an ongoing process of interaction and not a solidified pattern of exchange (Yeung, 1994). The ties that connect actors and nodes are in a continuous process of change. From time to time new sources of materials are discovered, new least cost centers emerge, trade preferences lose their market access values and so on and so forth. All these make network relationships unstable as old partners are forced to give ways to new and more qualified ones. To make such transitions easy with little adjustment costs, resident suppliers to global production networks have perfected a strategy of footloose engagement with their host economies. This they achieve by ensuring that their commitments to, or linkages with, their host communities are kept at the weakest possible form.

An unstable membership of a production network however, does not augur well for industrial development prospects in a region. Continuity is needed if sustainability of operations and hence, the goal of building a robust industrial base is to be achieved. To achieve this, it is required that the pattern of insertion that connects members with the activities of global production network actors must allow for the possibility of value upgrading (Kaplinsky and Morris, 2001). While it is possible to upgrade value through many different strategies, it is the practice of integrating backward that is especially useful for checking the enclave tendencies of resident suppliers to global production networks. Because it involves establishing linkages with local suppliers it has the potential to create a kind of local embeddedness that raises the stake against discontinuity of regional operations and relocation possibilities (see Na-Allah, 2009). Hence, a key challenge for any policy intervention like Systems of Trade Preferences directed at promoting national industrial development is to figure out how to incentivize the linkage development potentials of resident network firms.

5.3. Linkages and Systems of Trade Preferences

However, the idea of local inter-firm linkage formation is not totally strange to authors of system of trade preferences. In the course of its conception, it is feared that a real danger exists for non-beneficiaries to capture the gains of preferences through proxy trade that is made possible by superficial conditioning or assemblage of product items in beneficiary countries. As a result, a typical scheme is often packaged with built-in clauses to prevent trade deflection. A good example here is the Rules of Origin requirements or the so-called cummulation provision of many agreements. By stipulating that a certain percentage of production activities, relating to a particular product item, must be carried out in beneficiaries' economies, the rule or provision, as the case may be, is inadvertently encouraging the formation of production linkages in the area.

But beyond this anti-trade deflection function, it has been pointed out (mostly by critics) that due to their imposition of often-significant compliance capacity constraints on beneficiaries, the Rules have become a veritable instrument of creeping protection by donors. With much of the impact assessment studies becoming visible in this area, many findings have generally confirmed the claim (Olarreaga and Ozden 2005; Ozden and Sharma 2006; Carre re and de Melo 2006; Cadot and Melo, 2008)

But arguably, the most developmentally relevant justification for imposition of the Rules is that they can help foster the emergence of integrated manufacturing activities in beneficiary countries (Cadot and Melo, 2008). This claim aligns well with the spirit of linkage related value upgrading and general industrial development prospects of regions. It represents another dynamic area where impact assessment attention should be focused. Unfortunately, current research has done little to provide evidence on the extent of industrial linkage development attributable to the implementation of these trade programs. We can state the third and final aspect of our propositions as follows:

Emergence of integrated manufacturing activities through regional inter-firm linkages is the third dimension of potential effects of programs on industrial development and represents another aspect of dynamic impact issue.

6. SUMMARY

The discussion above reveals that analysis of the impact of system of trade preferences on industrial development outcome must proceed along at least three dimensions: export creation needed to upgrade scale operations; productivity improvement required for enhancing competitiveness and linkage development necessary for achieving sustainable development.

This three dimensional framework is captured in Figure 2. The top section illustrates the natural progression from trade preference shock to the three manifestations of development outcome: export stimulation, productivity upgrading and linkage development. Directly beneath this is a classification of the three outcomes in terms of static and dynamic impact. The combination of these two impacts is what produces the desired industrial development outcome.

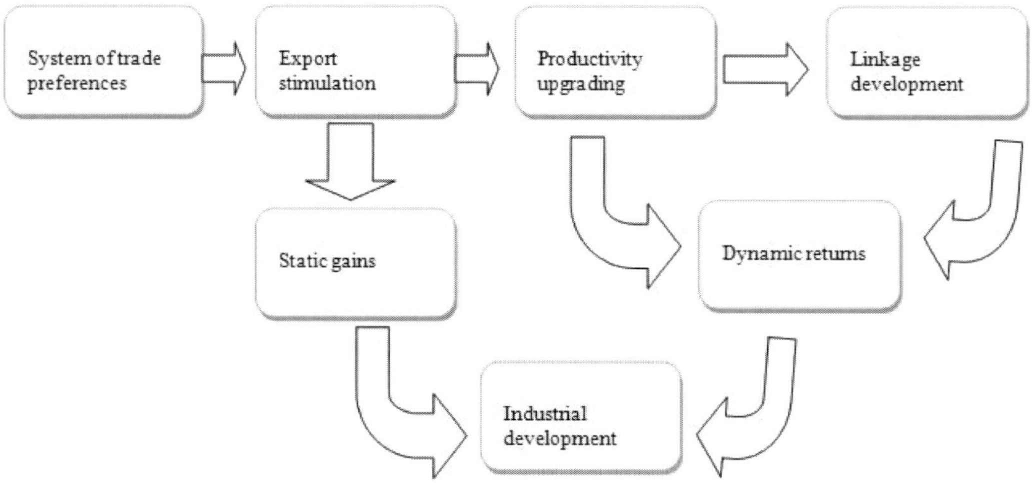

Figure 2. A vintage framework for analysis of the impact of trade preferences on industrial development.

CONCLUSION

In contemporary relations between developed and developing countries, the system of trade preferences for development has come a long way. Among the list of its designated objectives, promotion of industrial development arguably represents the most important. Despite this, extant literatures on the impact of program implementation have told only part of the story. We argue that this cannot be divorced from a lack of good appreciation of the multi-dimensional nature of the relationship between preferential trade incentives and the process of industrial development in beneficiary countries. For instance, while the goal of building a strong industrial base will require an intervention of the incentive schemes in some static and dynamic ways, current literatures focus only on the former.

In light of this, the chapter developed and presented a vintage framework, which recognizes the facilitator roles of the incentive packages in three main areas: export development (static impact), productivity improvement and emergence of linkage related integrated manufacturing (dynamic impact). This will hopefully help guide future research towards more comprehensive engagements.

REFERENCES

[1] Cadot, O., & Melo, J. D. (2008). Why OECD countries should reform Rules of Origin. *The World Bank Research Observer*, 23 (1): 77-105.

[2] Carre□re, C., & de Melo. J. (2006). Are different Rules of Origin equally costly? estimates from NAFTA. In O. Cadot (Ed.), *The Origin of Goods: Rules of Origin in Regional Trade Agreements.* London: Oxford University Press.

[3] Dickens, P. (2003). *Global Shift: Reshaping the Global Economic Map in the 21st Century*. 4th ed. London: Sage.

[4] Dickens, P. (2005). A New geo-economy. In D. Held & A. Mcgrew (Eds.), *The Global Transformations Reader.* UK: Polity Press.

[5] Gereffi, G. (2005). The global economy: organisation, governance, and development. In N. J. Smelser & R. Swedberg (Eds.), *The Handbook of Economic Sociology*.2nd Edition. New York: Sage.

[6] Held, D., Mcgrew, A., Goldblatt, D. & Perraton, J. (2005). Rethinking globalisation. In D. Held & A. Mcgrew (Eds.), The *Global Transformations Reader.* UK: Polity Press.

[7] Kaplinsky, R. & Morris, M. (2001). A handbook for value chain research, available from: http://www.ids.ac.uk/ids/global/pdfs/VchNov01.pdf.

[8] Lederman, D. & Ozdeñ, C. (2004). US trade preferences: all are not created equal, available from: http://wbln0018.worldbank.org/lac/lacinfoclient.nsf/;

[9] Mattoo, A. Roy, D. & Subramanian, A. (2002). The African Growth and Opportunity Act and its Rules of Origin: generosity undermined? IMF Working Paper WP/02/158

[10] Modeski, G. (2005). Globalisation. In D. Held & A. Mcgrew (Eds.), *The Global Transformations Reader.* UK: Polity Press.

[11] Na-Allah, A. (2009). Determinants of industrial embeddedness: evidence from African manufacturing firms. UNU-WIDER Working Paper wp2011-008. Available from: http://www.wider.unu.edu/ publications/working-papers/2011/en_GB/wp2011-008/

[12] Nilsson, L. (2002). Trading relations: is the road map from Lome to Cotonou correct? *Applied Economics,* 34: 439-452.

[13] Olarreaga, M. & Ozden, C. (2005). AGOA and apparel: who captures the tariff rent in the presence of preferential market access? *World Economy* 28(1):63–77.

[14] Onguglo, B. F. (1999). Developing countries and unilateral trade preferences in the new international trading system. In R. M. Miguel, L. Patrick & K. Barbara (Eds.), *Trade Rules in the Making: Challenges in Regional and Multilateral Negotiations.* Washington, D.C.: The Brookings Institution Press/Organization of American States

[15] Ozden, C., & Sharma, G. (2006). The price effects of preferential market access: Caribbean Basin Initiative and the apparel sector. *World Bank Economic Review* 20(2):241–60.

[16] Panagariya, A. (2000). Preferential trade liberalization: the traditional theory and new developments. *Journal of Economic Literature.* vol; xxxviii; 287-331.

[17] Rolfe, R. J. & Woodward, D. P. (2005). African apparel exports, AGOA, and the trade preference illusion. *Global Economy Journal*. 5 (3): 1-26.

[18] Sapir, A. & Ldunberg, L. (1984). The US Generalized System of Preferences. In R. Baldwin & Krueger (Eds). *The Structure and Evolution of Recent US Trade Policy*. Chicago: NBER.

[19] Seyoum, B. (2006). US trade preferences and export performance of developing countries: evidence from the generalised system of preferences. *International Business Review*. 15, 68-83.

[20] UNCTAD 1985). The History of UNCTAD 1964-1984 (UNCTAD/OSG/286), United Nations, New York (Sales No.E.85.II.D.6).

[21] WTO. (2001). Doha Ministerial Declaration. Available from: http//www.wto.org.

[22] WTO. (2005). Understanding the WTO. Available from: http//www.wto.org.

[23] Yeung, H.W. (1994). Critical reviews of geographical perspectives on business organisations and the organisation of production: towards a network approach. *Progress in Human Geography*. 18(4): 460-490.

*Reviewed by:*Sanya Osha (PhD). Senior Research Fellow, Institute for Economics Research on Innovation, TUT, Pretoria, South Africa. E-mail: oshaoo@tut.ac.za

In: Economics of Innovation, Incentives and Uncertainty
Editors: P. Simmons and S. T. Jordan
ISBN: 978-1-62257-251-9

Chapter 6

THE INNOVATION AND EXPORT ACTIVITY CAUSAL LINK: A REVIEW OF EXTANT LITERATURE

***Joaquín Monreal-Pérez*[1] *and Gregorio Sánchez-Marin*[2]**
[1]Assistant Professor at the Department of Management and Finance of the University of Murcia (Spain) and Research Affiliate at the Institute for the World Economy, Kiel, Germany
[2]Associate Professor at the Department of Management and Finance of the University of Murcia, Spain

ABSTRACT

In this work, we review the literature on the link between innovation and exports. In particular, we try to study the causal direction of the relation between both variables: it is whether innovation improves the export activity of the firm, as suggested by the self-selection theory, or otherwise, whether it is the greater firm export activity what enhances the innovative performance of the firm, as predicted according to the learning-by-exporting arguments. We focus on a number of recent studies that, dealing with the endogeneity concern, show stronger support for the self-selection theory on innovation, rather than for the reverse link, stated by the learning-by-exporting hypothesis. This conclusion matches up with the most common finding of the export literature when considering the firm productivity.

1. INTRODUCTION

The study of the export activity of the firm is of great opportunity, especially when considering that due to the current economic crisis, quite sound in most of the developed markets, selling the goods or services only in the domestic markets is becoming increasingly more difficult for the firm. Moreover, exporting provides a number of advantages for the firm: it improves the firm's financial position, it helps to develop the firm competitive advantages, it enriches the managerial skills, it makes a better utilization of the production capacity, and it facilitates the growth of the firm (Leonidou *et al*., 2007).

In this context, the literature has made an effort to answer two questions, depicting a double perspective of the same matter: first, which are the drivers of the firm export activity, it is, which are the firm resources or capabilities whose possession implies that the export markets *select* them, as the self-selection hypothesis advocates try to explain?; and, second, apart from the commonly accepted export activity benefits for the firm, does such activity exert any positive impact on the firm performance? In this sense, the learning-by-exporting hypothesis argues that the export activity will favor that the firm (thanks to the knowledge acquired while exporting, and to the fierce competition within the export markets), perform better than when operating only in its market of origin.

The empirical findings of the literature on these two explanations offer a much stronger support for the self-selection hypothesis, rather than for the learning-by-exporting one (Wagner, 2007). Nevertheless, the results of most of these studies may be biased due to the fact that they do not focus on both theories jointly, especially when analyzing the innovative behavior of the firm. Thus, considering the compatibility within both theories (Love and Mansury, 2009), we wonder if it is the self-selection or the learning-by-exporting theory which explain robustly better the relationship between innovation and exporting, at a micro level?

Among the main determinants of the firm export competitiveness, a number of studies of the export literature have underlined the great strategic importance of the firm innovation in the export markets (Leonidou *et al*., 2007). Additionally, the innovative activities strategic significance is due to the fact that through such activities, the firm can differentiate itself from its rivals within the export market (López and García, 2005; Filipescu *et al*., 2009; Pla-Barber and Alegre, 2009).

In short, taking into account the arguments stated above, some recent works that disentangle the relationship between the firm export behavior and its innovation activity are reviewed. Specifically, two concerns are addressed: first, it is studied the impact of these variables that, due to their significance when studying the firm performance (eg. productivity), may moderate these relationships; and, second, most of the works focus on one of the two hypotheses (the self-selection and the learning-by exporting one) on a isolate basis. Nevertheless, there may be an endogeneity problem, it is, the dependent variable of the model (the export activity one, in the self-selection model or the innovation covariate in the learning one) may influence the independent variables, what suggests the presence of the opposite effect. Thus, we will focus on the works addressing the endogeneity, and, especially, on the way in which this matter has been faced.

2. Innovation and Exports: Coexistence of Two Hypotheses

The important role of being innovative to become more competitive internationally has been underlined in a number studies. Thus, Leonidou *et al*. (2007) argue that the development of a special innovative capacity is one of the principal stimuli that drive a firm to engage in export activities. Moreover, innovation can differentiate a firm and thereby serve as a source of international competitive advantage (López and García, 2005; Filipescu et al., 2009; Pla-Barber and Alegre, 2009).

Besides, from a resource based perspective (Wernerfelt, 1984), innovation serves as a source of sustainable competitive advantage mainly because innovation is inimitable. A firm innovative capacity represents an original combination of the organization resources and is developed over the lifetime of the firm. Imitation is difficult because the firm's competitors may not possess the necessary resources to fully exploit this capacity (Miller and Shamsie, 1996). Pla-Barber and Alegre (2007) stress the role of innovation in international markets by arguing that a single market may not be broad enough to support the innovation needs of the firm. As a result, an innovative firm may try to export instead. Moreover, innovative firms have incentives to expand their activities into other markets, as these firms can earn higher returns from their investments in other markets (Teece, 1996). In this context, the export markets may represent an area in which the firm can exploit their innovations and thereby enhance its economic performance (Love and Mansury, 2009; Filipescu *et al.*, 2009).

In spite of this generally accepted argument, there are two views in the way that the firm export activity and its innovation performance are causally related. These two explanations are described next.

a. Self-Selection Hypothesis

The self-selection effect related to the firm innovative activity implies that such activity intensifies the firm export behavior. This theory main arguments lay on: first, innovation can differentiate a firm and, therefore, serving as a source of international competitive advantage for the firm (López and García, 2005; Filipescu *et al.*, 2009; Pla-Barber and Alegre, 2007); second, through internationalization, the firm may access a wider consumer base, what makes it to be able to better exploit its innovation outputs (Love and Mansury, 2009).

The third argument is that the innovative activity implies a range of costs, what makes that consumers demand expansion reached through the export venture will reduce the firm unitary costs (Kotabe *et al.* 2002). Therefore, it is the tradeoff between the costs and the gains associated to the innovative activity what determines its profitability (Caldera, 2010). That is the reason why those more effective firms have enough margins to innovate. In this sense, García-Vega and López (2010) argue that financial resources destined for the firm innovative activity is the main perceived obstacle that the firm needs to face when willing to carry out innovations.

Confirming empirically these arguments, a number of studies have shown that the innovative activity acts as an important driver of exports (Filipescu *et al.*, 2009; Harris and Li, 2009), what suggests that causality runs from undertaking innovation activities to internationalization. The literature on innovation management generally predicts that innovative firms will enter a foreign market to increase their sales volume (Tidd *et al.*, 1997). Hanley and Monreal-Pérez (2012) find that the *ex ante* most productive and high-tech firms self-select into export markets. Additionally, when focusing on the firm innovative activity, these authors state that a firm Research and Development (RandD) investments are positively related to export entry.

Moreover, if firms adjust their products to foreign customer needs and reduce their operational costs (through process innovations), they stand to gain higher export shares in these new markets (Lachenmaier and Wößmann, 2006). Thus, it can be argued that according to this self-selection theory, the more innovative firms obtain better results in the export markets. Then, the endogeneity concern affecting such hypothesis implies that the innovation proxies may be influenced by the export activity variables (what suggest the reverse explanation, it is, the learning-by-exporting one, effect that is explained next).

b. Learning-by-Exporting Hypothesis

On the other hand, the greater competitive pressure of the export markets (Wagner, 2007) and the knowledge acquired in these markets (Nonaka and Takeuchi, 1995; Hitt et al., 1997; Kotabe et al., 2002; Alegre and Chiva, 2008) are the two most important arguments supported by the learning-by exporting advocates. In this sense, Girma *et al.* (2003) observe that the benefits from exporting are due to exposure to best practice technology. Thus, exporters stand to make productivity and other performance indicators' gains (eg. in the firm innovative behavior), as their costs get spread over expanded output, as happens when exporting.

Several papers have shown that a positive relationship exists between a firm's exposure to foreign markets and the level and growth of the firm technological knowledge (e.g., Alvarez and Robertson (2004) for Mexican and Chilean enterprises; Cassiman and Veugelers (1999) for Belgian firms). Scholars generally agree that this positive connection is caused by the highly competitive pressure of international markets, which forces firms to constantly update their products and adapt to new market conditions.

In the same way, Salomon and Shaver (2005), when conducting a case study of Spanish firms, find that the knowledge acquired in the export market allows a firm to register more patents and develop more innovative products. These authors point out the importance of exploring the time path of the learning-by-exporting effect, as the impact on innovation may not be immediate. They find that the learning-by-exporting effect primarily affects product innovation two years after the firm begins exporting and that the number of patents registered increases with a greater lag in time.

Consistently with Bustos (2011), Hanley and Monreal-Pérez (2012) find a significant correlation between the switch to exporting and the appearance of newly introduced products / manufacturing processes contemporaneous with the switch. Moreover, these authors also find that the positive effect carries through the year following the transition, $t+1$. Following, Hanley and Monreal-Pérez (2012), applying a combination of propensity score matching with difference-in-differences to a cohort of newly exporting Spanish firms in 2006, state that little evidence of returns to a firm innovative activity from exporting, the exception being an 11 percent increase in the rate of process innovations for newly exporting firms in the year that they transition to exporting.

Nevertheless, in a surprising conclusion, Silva and Leitão (2007) state that Portuguese industrial firms with high export intensity were less capable of innovating. They argued that the majority of high export intensity firms worked on an outsourcing basis, adopting a low-price strategy which did not rely on product innovation.

However, the results may have been biased due to the fact that most of the results have been carried out focusing on the productivity rather than in the innovative activity, and especially from a macroeconomic view or, when using a microeconomic perspective, on big firms (mainly MNEs), rather than in small ones.

Thus, although empirical evidence appears to support more the self-selection hypothesis than the learning-by-exporting one, a number of studies show that these two explanations are complementary and not mutually exclusive (Wagner, 2007; García and Avella, 2008). In other words, the markets *select* better performing firms, at the same time as the exposure to international markets makes these firms become even better. Accordingly, some scholars have found evidence suggesting that the export activity leads to higher productivity even after accounting for the self-selection effect (Van Biesebroeck, 2005; Love and Mansury, 2009).

Finally, as Love and Mansury (2009) note, although these two scenarios may coexist, from a public policy and a firm strategy perspective, one must know which one better explains the causal relationship between innovation and exports, since the identification of whether only one scenario or both scenarios hold in practice implies the adoption of different policies.

3. Additional Research Exploring the Innovation-Exports Causal Links

a. Moderating Determinants

Studying the factors moderating the self-selection and the learning-by-exporting hypotheses will allow managers and policymakers to promote the determinants that can strengthen the benefits of the firm productivity, innovation and the export activities. Once that the importance of analyzing such moderators has been pointed out, the next question that needs to be answered is: which are the factors with such moderating capability?

Monreal-Pérez *et al.* (2011) underline that the firm productivity may play a significant role due to the following reasons: first, and considering the self-selection hypothesis, those more productive firms are more likely to export because the level of competition in the export markets is more intense, as these markets have higher costs than the domestic markets (Roberts and Tybout, 1997; Clerides *et al.*, 1998, and Helpman *et al.*, 2004). Second, the learning-by-exporting hypothesis states that firms benefit from the knowledge generated by exporting to international markets (García and Avella, 2008; Delgado *et al.*, 2002). In turn, this additional knowledge renders the firms more productive and improves their development processes once they start exporting their products.

According to the results from Monreal-Pérez *et al.* (2011), none of the interactions between productivity and innovation are significant, what suggests that policymakers have an incentive to implement measures that aim to increase both innovation and productivity, but the joint effect of both variables is not necessary to succeed in the export markets.

On the other hand, Monreal-Pérez *et al.* (2012) stress the international mode of entry as a potential moderator of the innovation-exports relationship. In line with the transaction costs theory prescriptions (Teece, 1986), the authors conclude that those innovative firms who adopt international export modes of entry implying high levels of control (for instance, exporting directly through a commercial distributor belonging to the firm and based in the country of destination of the export), develop greater export intensity than those exporting through an indirect entry mode.

b. Endogeneity Concerns

Nevertheless, as Salomon and Shaver (2005) note, what is really most important is finding the temporal sequence that drives the relationship between innovation and exports to draw conclusions about the firm export strategy.

In this sense, the empirical literature on exports has become increasingly aware of the need for disentangling the direction of causality between exports and several measures of firm performance, among which the firm innovation behavior stands out.

The studies facing this concern related to the innovation-exports literature can be classified into two different categories: those dealing with the potential presence of an endogeneity problem affecting the self-selection theory, and those on the endogeneity of the learning-by-exporting hypothesis.

Regarding the first group, the primary concern is the identification of a good instrument for the innovation output. Caldera (2010) argues that the Public Subsidies for the firm's RandD expenditures (PSRD) may be a good instrument for measuring innovation because these public policies aim to increase the firm innovative activities and are unrelated to the firm export behavior. To test for the validity of PSRD as an instrument, Monreal-Pérez *et al.* (2011), trough the correspondent correlation tests, confirm that this variable fulfils both the inclusion restriction (i.e., the instrument must be correlated with the endogenous variable) and the exclusion restriction (i.e., the instrument cannot be correlated with the error term in the model). Thus, these two studies using this find that the results of the basic models (showing that innovating improves the firm's export propensity) are robust to endogeneity.

Lachenmaier and Wößmann (2006), using a set of innovation impulses and obstacles as instruments, argue that innovation leads to an increase of seven percentage points in the export share of German manufacturing firms. These authors distinguish between the two main innovation outputs: the product and the process innovation. When instrumenting these variables, the impulses stemming from the firms' Production and Resource Management Department lead to more product innovation, while the obstacles caused by the lack of equity capital reduce both types of innovations.

Besides, taking into account that the past export status of the firm is a key explanatory variable capturing the sunk cost effect of the previous investments in exporting (Roberts and Tybout, 1997), Caldera (2010) and Monreal-Pérez *et al.* (2011) consider the multi-period case including the lagged outcome. Thus, these scholars build an Arellano and Bond's (1991) first-difference GMM estimator, finding that the export markets *select*, at a greater extent, those firms developing an intense innovative activity, rather than the non-innovative ones.

Cassiman and Martínez-Ros (2004) lag the innovation proxies to argue that innovations are endogenous and, therefore, their exogenous treatment leads to largely downward-biased estimates of the impact of innovations on firm-level exports. As instruments for product and process innovation, Cassiman and Martínez-Ros (2004) use industry and time dummies together with the percentage of firms that obtained a product or process innovation, respectively, in that particular industry for that particular year. These authors conclude that the gathering of product innovations exerts a great self-selection effect, since it increases the firm export activity.

Regarding the second group of studies (those analyzing the endogeneity of the self-selection hypothesis), the main matter to address when following an Instrumental Variables (IV) strategy is again finding a valid instrument.

To solve this matter, Bratti and Felice (2010) point out that a firm export status should be strongly negative correlated with the distance between its geographical location and the potential destination for its products, as transportation costs generally increase with distance. Monreal-Pérez *et al.* (2011) test that this variable is valid since it fulfills the inclusion and the exclusion restriction. Thus, Bratti and Felice (2010) and Monreal-Pérez *et al.* (2011) use the export distance as an instrument, calculated according to the great circle distances procedure.

Although Bratti and Felice (2010) find that exporting has a large positive effect on the likelihood of introducing product innovations, according to Monreal-Pérez *et al.* (2011), the residuals obtained from the regression are significant in the Ordinary Least Squares (OLS) basic model, but not significant for neither the product innovation, nor the process innovation variable, what suggests that the firm export propensity does not show a endogeneity problem. Similarly, Bratti and Felice (2010) also conclude that export activity was exogenous to the learning effect exerted by this variable on the firm's innovative capacity when controlling for the firm labor productivity. Because of that, the two mentioned works considered the export activity covariates as exogenous and, therefore, did not carry out any strategy to deal with the endogeneity robustness of the results associated to the learning-by-exporting explanation.

Conclusion

Although the self-selection and the learning-by-exporting hypotheses have been deeply analyzed and reviewed when considering the firm productivity (as in Wagner, 2007), the studies dealing with the link between innovation and the firm export behavior are scarce. This gap is greater when willing to address the potential endogeneity concerns on both explanations. This study reviews a number of recent studies which show that the self-selection hypothesis offers a better explanation of the analyzed relationship between innovation-exports than the learning-by-exporting one. This result is robust to endogeneity, problem faced in different ways (lagging the exogenous variables, propensity score matching, using instrumental variables, through GMM estimators, ...).

Among the different innovation variables used, the product innovations showed to matter the most, especially at the prior moment in which the firm are self-selected by the export markets (Lachenmaier and Wößmann, 2006; Cassiman and Martínez-Ros, 2004; Becker and Egger, 2007), since when product innovating, the firm differentiate itself from its competitors and, therefore, gain a competitive advantage that is highly appreciated within the export markets. Following, if the firm wants to keep its competitive position in the market, becoming more productive (for instance, through process innovations) is of great importance. Thus, exporters may have to align production to help meet the myriad needs of overseas markets and reduce costs (Lachenmaier and Wößmann, 2006).

However, lot of research in this field needs to be done: the literature offering robust results is scant, since most of the findings are biased because of the fact that they don't consider the potential endogeneity between the variables. Moreover, we also recommend studying how a number of determinants moderate the hypotheses explaining the innovation-exports link. In this sense, as far as we know, only the moderating effect of the mode of entry and of the firm productivity has been considered (Monreal-Pérez *et al.* 2011, 2012).

Besides, since this work focuses on only one international entry mode, the export one, we propose analyzing how the effects studied differ when considering others internationalization strategies, as the Foreign Direct Investment (FDI) or the alliances one. The need for widening the study to other modes apart from exporting is important, especially when analyzing the learning by exporting effect. This is due to the fact that, although exporting facilitates the flow of information from the host market, exporting does not provide sufficient information to generate the predicted effects. In these circumstances, more involved methods of international expansion (e.g., FDI) are required to generate knowledge about the local environment (Salomon and Shaver, 2005). This argument leads to Salomon and Shaver (2005) to argue that scholars have failed to find evidence in favor of the learning-by-exporting effect because they have neglected the method of international expansion. In this sense, Kafouros *et al.* (2008) point out that firms need to achieve a minimum degree of internationalization to appropriate the fruits of innovation. This minimum level is reached preferably through entry modes involving a higher commitment, as in FDI, rather than when the firm exports.

Finally, it is important to consider that, both the self-selection theory and, especially, the learning-by-exporting effect, becomes more evident when the export comes from a less-developed country to a relative more developed one. Thus, (apart from recent efforts, as in Bratti and Felice, 2010; or in Monreal-Pérez *et al.*, 2011, who control for the export distance, but just for an instrumental purpose), it would be interesting to carry out these analyses on countries in which the exports goes to a significantly more developed market (Silva *et al.*, 2010). This strategy was followed by Salomon and Shaver (2005), who find a strong support for the learning by exporting effect when analyzing a sample of exports coming from manufacturing firms belonging to a number of sub-Saharan African countries. Nevertheless, it is recommended to develop this type of analyses, but considering both the moderating and the endogeneity concern affecting potentially these two effects.

REFERENCES

Alegre, J. and Chiva, R. (2008) Assessing the impact of organizational learning capability on product innovation performance: An empirical test, *Technovation*, 28, 315– 326.

Alvarez, R and Robertson, R. (2004). Exposure to foreign markets and plant-level innovation: evidence from Chile and Mexico. *The Journal of International Trade and Economic Development,* 13(1), 57-87.

Bratti, M. and Felice, G. (2010). Are exporters more likely to introduce product innovations?, *EFIGE working paper*, 25, 1-36.

Bustos, P. (2011). Trade Liberalization, Exports, and Technology Upgrading: Evidence on the Impact of MERCOSUR on Argentinian Firms, *American Economic Review*, 101, 304–340.

Caldera, A. (2010). Innovation and exporting: evidence from Spanish manufacturing firms. *Review of the World Economics*, 146, 657–689.

Cassiman, B. and Martínez-Ros (2004). Innovation and Exports: Evidence from Spanish Manufacturing, IESE Business School, Unpublished manuscript, 1-36.

Cassiman, B. and Veugelers, R. (1999). Make and buy in innovation strategies: evidence from Belgian manufacturing firms. *Research Policy*, 28, 63-80.

Filipescu, D. A., Rialp, A. and Rialp, J. (2009). Internationalisation and technological innovation: empirical evidence on their mutual relationship. *Advances in International Marketing*, 20, 125-154.

García, F. and Avella, L. (2008). La influencia de la exportación sobre los resultados empresariales: análisis de las pymes manufactureras españolas en el período 1990-2002. *Revista Europea de Dirección y Economía de la Empresa*, 17(2), 85-104.

García-Vega, M. and López, A. (2010). Determinants of Abandoning Innovative Activities: Evidence from Spanish Firms, *Cuadernos de Economía y Dirección de Empresa*, 45, 69-91.

Girma, S., Greenaway, D., Kneller, R., 2003. Export market exit and performance dynamics: a causality analysis of matched firms. *Economics Letters*, 80 (2), 181–187.

Hanley, A. and Monreal-Pérez, J. (2012). Are Newly Exporting Firms more Innovative? Findings from Matched Spanish Innovators. *Economics Letters,* doi:10.1016/ j. conlet.2012.03.006.

Harris, R., and Li, Q. C. (2009). Exporting, RandD, and absorptive capacity in UK establishments. *Oxford Economic Papers*, 61(1), 74–103.

Hitt, M. A., Hoskisson, R. E. and Kim, H. (1997). International diversification: effects on innovation and firm performance in product-diversified firms. *Academy of Management Journal*, 40(4), 767-798.

Kafouros, M. I., Buckley, P. J., Sharp, J. A. and Wang, C. (2008). The role of internationalization in explaining innovation performance. *Technovation*, 28(1/2), 63-74.

Kotabe, M., Srinivasan, S. S. and Aulakh, P. S. (2002). Multinationality and Firm Performance: The Moderating Role of RandD and Marketing Capabilities. *Journal of International Business Studies,* 33(1), 79-97.

Lachenmaier, S. and Wößmann, L. (2006). Does innovation cause exports? Evidence from exogenous innovation impulses and obstacles using German micro data. *Oxford Economic Papers*, 58(2), 317-350.

Leonidou, L. C.; Katsikeas C. S.; Palihawadana, D.and Spyropoulou, S. (2007). An analytical review of the factors stimulating smaller firms to export. Implications for policy makers. *International Marketing Review*, 24(6), 735-770.

López, J. and García, R.M. (2005). Technology and Export Behaviour: A Resource-Based View Approach. *International Business Review*, 14(5), 539-557.

Love, J.H. and Mansury. M.A. (2009). Exporting and productivity in business services: Evidence from the United States. *International Business Review*, 18, 630–642.

Miller, D. and Shamsie, J. (1996). The Resouce-Based View of the Firm in two Environments: The Hollywood Film Studios from 1936-1965. *Academy of Management Journal,* 39(1), 519-543.

Monreal-Pérez, J.; Aragón-Sánchez, A. and Sánchez-Marín, G. (2012). Innovation and the firm export intensity: the moderating effect of the mode of entry. *Esic Market*, 141, 27-46.

Monreal-Perez, J.; Aragón-Sánchez, A. and Sánchez-Marín, G. (2011). A longitudinal study of the relationship between export activity and innovation in the Spanish firm: The moderating role of productivity. *International Business Review.* Article in press.

Nonaka, I. andTakeuchi, H. (1995). The knowledge-creating company. Oxford University Press, New York.

Pla-Barber, J.and Alegre, J. (2007). Analysing the link between export intensity, innovation and firm size in a science-based industry. *International Business Review,* 16, 275-293.

Roberts, M. and Tybout, J. (1997). What Makes Exports Boom, The World Bank, Washington DC.

Roper, S. and Love, J.H. (2002). Innovation and export performance: evidence from the UK and German manufacturing plants. *Research Policy,* 31(7), 1087-1102.

Salomon, R.M.; and Shaver, J. M. (2005). Learning by Exporting: New Insights from Examining Firm Innovation. *Journal of Economics and Management Strategy,* 14(2), 432-460.

Silva, A, Africano, A.P. and Afonso, O. (2010). Do Portuguese manufacturing firms learn by exporting? 36th *EIBA Conference Proceedings*, Porto, 2010.

Silva, M.J. and Leitão, J. (2007). What determines the entrepreneurial innovative capability of Portuguese industrial firms?. *Munich Personal RePEc Archive Paper*, n. 5216.

Teece, D.J. (1986). Transaction Cost Economics and the Multinational Enterprise: An Assessment. *Journal of Economic Behavior and Organization*, 7, 21-45.

Teece, D.J. (1996). Firm organization, industrial structure and technological innovation. *Journal of Economic behavior and Organization*, 31, 192–224.

Tidd, J., Bessant, J., and Pavitt, K. (1997). Managing innovation: Integrating technological, market and organizational change. Chichester: John Wiley and Sons.

Van Biesebroeck, J. (2005). Exporting raises productivity in sub-Saharan African manufacturing firms. *Journal of International Economics*, 67, 373–391.

Wagner, J. (2007). Exports and productivity: a survey of the evidence from firm-level data. *The World Economy*, 30(1), 60-82.

Wernerfelt, B. (1984). A Resouce-Based View of the Firm. *Strategic Management Journal,* 5, 171-180.

In: Economics of Innovation, Incentives and Uncertainty ISBN: 978-1-62257-251-9
Editors: P. Simmons and S. T. Jordan

Chapter 7

MEASURES OF UNCERTAINTY IN ECONOMICS

Mohsen Bahmani-Oskooee[1,*] and Scott W. Hegerty[2]

[1]Center for Research on International Economics,
University of Wisconsin-Milwaukee, Milwaukee, WI, US
[2]Department of Economics, Northeastern Illinois University,
Chicago, IL, US

ABSTRACT

Uncertainty in labor and financial markets has received a great deal of blame for the United States' recent sluggish macroeconomic performance. As a result, it can be expected that future studies will focus on the role of this risk in determining a wide variety of macroeconomic variables. How do we measure uncertainty in economics? In this chapter we review different measures of uncertainty that are widely applied in the literature, focusing on those that are based on variance, prediction error, and GARCH modeling. For demonstrative purposes, we generate these different measures of uncertainty for the monthly dollar-pound real exchange rate and compare their statistical properties. In general, all of our measures capture the same periods of high volatility, but their statistical properties vary widely.

JEL Codes: B4, C2

Keywords: Uncertainty, Volatility, Time Series, Exchange Rates

I. INTRODUCTION

In an era of renewed macroeconomic instability, understanding these fluctuations has taken on added importance. It is therefore crucial that economists accurately measure this uncertainty. While the end of the "Great Moderation" in 2008 came as a surprise to some, there has been a long tradition of devising and implementing empirical proxies for

[*] E-mail: bahmani@uwm.edu.

uncertainty. Over the past few decades, various measures of economic volatility have been used to study numerous topics. Examples include real variables such as output and its components (Ramey and Ramey, 1995; Fang and Miller, 2008; Ahmed and Suardi, 2009), as well as capital flows (Broner and Rigobon, 2004), interest rates (Edwards, 1998; Laopodis 2000, 2001, 2002; Edwards and Susmel, 2003), and exchange rates (Sauer and Bohara, 2001; Bahmani-Oskooee and Hegerty, 2009a,b), among many others. No single measure of volatility has gained wide enough acceptance to dominate the others. The fact that the researcher is faced with a number of choices of volatility proxies has enriched the literature, but it makes it more difficult to ascertain which method is preferred, and how related alternative measures might be.

The set of available choices differs as well, depending on the nature of the data that are being studied. While cross-sectional studies can make use of multi-year averages that do not need to be recalculated year after year, time-series analyses require a continuous, time-varying proxy for volatility that does not introduce obvious problems such as autocorrelation to the series. These therefore are of particular concern. In this study, we examine a set of possible alternative measurements for macroeconomic uncertainty. Drawing upon the literature review by Bahmani-Oskooee and Hegerty (2007), we focus on one branch of the literature—exchange-rate volatility—which has seen many of these implemented in practice. We highlight the major methods of calculating time-series volatility proxies before using each of these to calculate the variability of the pound-dollar real exchange rate. We find that these methods are not always highly correlated and that some are more prone to persistence and autocorrelation than are others.

II. Major Measures of Time-Varying Volatility

First, we must make clear that "volatility" and "uncertainty" are not perfect synonyms, although the literature often treats them as such. In particular, our main time-series measures approach these proxies in two different ways. In one, the variation of each series is calculated. In the other, an "unknown" component is extracted from a predictive model, representing the underlying risk associated with the process. Therefore, of the many methods of calculating exchange-rate volatility, we focus on three major families of measure that are derived from these two concepts. These include standard deviations (both within-period or moving), measures based on prediction error, and ones that apply various forms of Autoregressive Conditional Heteroskedasticity (ARCH) models. It is important to note that this review is not exhaustive: We do not focus on certain other measures that rely on outside data sources, for example, such as ones that incorporate measures of investor expectations when calculating uncertainty. Nor do we delve into more advanced methods, such as stochastic volatility modeling. We describe each of our main measures in turn.

II.1. Measures Based on Variance

Probably the most common method, particularly when using short or low-frequency datasets, is the standard deviation. This allows for an accurate measurement of a series' variance, which serves as the basis of these volatility measures. These can divided into two sub-groups: Standard deviations based on higher-frequency subsamples within a given time period, and those that are calculated over rolling windows. In the exchange-rate volatility literature, this within-period standard deviation is calculated as

$$VOL_t = \sqrt{\frac{1}{n-1}\sum_{i=1}^{n}\left(E_i - \overline{E}\right)^2} \tag{1}$$

where E is either a nominal or real exchange rate and n is generally 12. This measure is used in the seminal paper by Akhtar and Hilton (1984), who calculate an annual volatility term using the standard deviation of the 12 monthly exchange rate observations within each year. More recently, this approach has been applied by Bahmani-Oskooee and Hegerty (2009a) and Bahmani-Oskooee *et al.* (2010, 2011). While these papers generally base this measure on the *levels* of the exchange rate, this has been criticized for allowing exchange-rate volatility to depend partially on the strength of the currency (since higher means can lead to higher variance). As a result, studies such as Bahmani-Oskooee and Hegerty (2009b), for the case of the Japanese yen, use percentage changes of the underlying exchange rate.

Other papers use less common techniques to arrive at a similar concept. Bailey *et al.* (1987), for example, proxy volatility as absolute percentage changes in the exchange rate, or

$$\left|\frac{E_t - E_{t-1}}{E_{t-1}}\right| \tag{2}$$

A more frequently used version of these techniques involves calculating standard deviations over rolling windows. For each time period t,

$$VOL_t = \sqrt{\frac{1}{n}\sum_{i=1}^{n}\left(E_{t+i-1} - E_{t+i-2}\right)^2} \tag{3}$$

The size of the window n varies by study, but generally equals one or more years. For example, a quarterly study might have $n = 4$ or $= 8$, although some take the fact that earlier periods are included in the differencing and might have $n = 3$ or $n = 7$ for the length. One drawback to a longer window is that innovations are more persistent as the window lengthens, while short horizons suffer from small sample size. Major studies that have used this approach include those by Chowdhury (1993), Kenen and Rodrik (1986) and Sauer and Bohara (2001). We do so as well below.

II.2. Measures Based on Prediction Error

As noted previously, uncertainty can be measured by measuring the unpredictable component of a macroeconomic time series. If the underlying model proves unable of correctly forecasting the series in question, then risk is present. Likewise, forecasters face increased uncertainty for the same reasons if their predictions are inaccurate.

While a structural or reduced-form model may be appropriate as the basis of the underlying predictive model, simpler methods prevail. One of the simplest involves a simple regression of a quadratic trend model, with the exchange rate regressed on a time trend and this trend squared:

$$\ln E_t = \phi_0 + \phi_1 t + \phi_2 t^2 + \varepsilon_t \tag{4}$$

This accounts for both linear and nonlinear movements in the time series. The residuals (or standard errors, as explained below) can be used as a measure of uncertainty. These have been applied extensively in the literature. Papers that have used this method include De Grauwe (1987), Thursby and Thursby (1987), Grobar (1993), and Arize and Ghosh (1994). A more formal approach involves ARIMA modeling, but is used to achieve essentially the same goals:

$$\ln E_t = \alpha + \sum_{i=1}^{p} \rho_i E_{t-i} + \sum_{j=0}^{q} \theta_j \varepsilon_{t-j} \tag{5}$$

The appropriate autoregressive and moving-average lengths are chosen by the well-known Box-Jenkins technique. Once the model is specified, the residuals can be used to form a volatility proxy. Studies that employ this method include Asseery and Peel (1991), Grobar (1993) Arize and Shwiff (1998), and Sauer and Bohara (2001).

These models, particularly the latter one, can derive a proxy for uncertainty in by taking the absolute value of the residuals to measure the predictive error of the model. In an alternative specification, a simple AR(1) model can be estimated over rolling windows (of a pre-determined length), with the rolling standard error serving as a measure of uncertainty.

$$\ln E_t = \alpha + \rho_1 E_{t-1} + \varepsilon_t \tag{6}$$

In general, this method imposes a cost, since more observations are lost when longer windows are used. There are certain benefits, such as a shorter persistence after innovations, which we observe below when we compare all of our models side-by-side.

II.3. Models Based on Conditional Heteroskedasticity

While not nearly the most computationally advanced techniques available, methods that are based on volatility "clustering" have been widely applied in this branch of the literature.

In principle, since high volatility tend to occur in specific periods, disturbances might be shown to depend on their own past values. This forms the basis of Engel's (1982) ARCH modeling, where the error terms in a particular model follow some type of an autoregressive process:

$$\varepsilon_t = v_t \sqrt{\alpha_0 + \sum_{i=1}^{q} \alpha_i \varepsilon_{t-i}^2} \tag{7}$$

In this ARCH(q) model, all shocks to past error terms, modified by a white-noise term *v*, contribute to the current error. Thus, periods of high volatility will tend to cluster together. This type of modeling was extended by Bollerslev (1986) into the GARCH (Generalized ARCH) process, which includes past volatility measures as well as past error terms into the current volatility measure. While the order of a GARCH(*p,q*) model is flexible in theory, most empirical papers employ GARCH(1,1) processes:

$$\varepsilon_t = v_t \sqrt{h_t} \tag{8a}$$

$$h_t = \alpha_0 + \alpha_1 \varepsilon_{t-1}^2 + \beta_1 h_{t-1} \tag{8b}$$

In recent years, a large numbers of GARCH variations have been introduced. These include Integrated GARCH (IGARCH); Threshold GARCH (TGARCH), in which the model parameters may differ above and below a certain value; and Nonlinear GARCH (NGARCH). Here, we focus on two variations that may be particularly useful in the specific case of exchange-rate volatility: GARCH-in-Mean and Exponential GARCH. The former, introduced by Engle *et al.* (1987), incorporates the volatility into the mean equation itself. In our case, perhaps investors are wary of exchange-rate uncertainty and the resulting impact on investment weakens the currency:

$$\Delta y_t = c + \gamma\sigma^2 + \sum_{i=1}^{p} \rho_i \Delta y_{t-1} + \sum_{j=0}^{q} \theta_j \varepsilon_{t-j} \tag{9}$$

In the second variation, by Nelson (1991), positive and negative variations have asymmetric effects. Again, this might have a specific application for exchange-rate uncertainty, since investors and currency traders feel differently about rising currencies than they do about falling ones:

$$\ln(h_t) = \alpha_0 + \alpha_1 \left(\frac{\varepsilon_{t-1}}{h_{t-1}^{0.5}}\right) + \lambda_1 \left|\left(\frac{\varepsilon_{t-1}}{h_{t-1}^{0.5}}\right)\right| + \beta_1 \ln(h_{t-1}) \tag{10}$$

In this section of our study, we thus focus on GARCH(1,1), EGARCH(1,1), and GARCH-M(1,1) for both. Having outlined three branches of volatility measures, the next step is to construct them and compare their properties. This is done in the next section.

III. Application: The USD/GBP Real Exchange Rate

How similar are these various measures of exchange-rate uncertainty? In this section, we construct a number of these various alternative uncertainty measures for a time series of the pound-dollar real exchange rate. Data, taken from the International Financial Statistics of the International Monetary Fund, are monthly and cover the period from January 1960 to November 2011 (623 observations). The nominal exchange rate is in terms of pounds per dollar; the countries' Producer Price Indices (2005 = 100) are used to construct the real rate. We see this exchange rate, both in levels and first differences, in Figure 1.

Clearly, exchange-rate volatility increases after the 1973 breakdown of the Bretton Woods agreement. Periods during which large exchange-rate movements occur include the 1967 pound devaluation, the period after the Plaza Accord in the 1980s, the European Monetary System's crisis of the early 1990s, and the Global Financial Crisis of 2008. Large swings are not automatically synonymous with periods of exchange-rate volatility, but we do find these periods when we create our uncertainty indices, and use these as our benchmark for whether a particular measure is capable of capturing uncertainty.

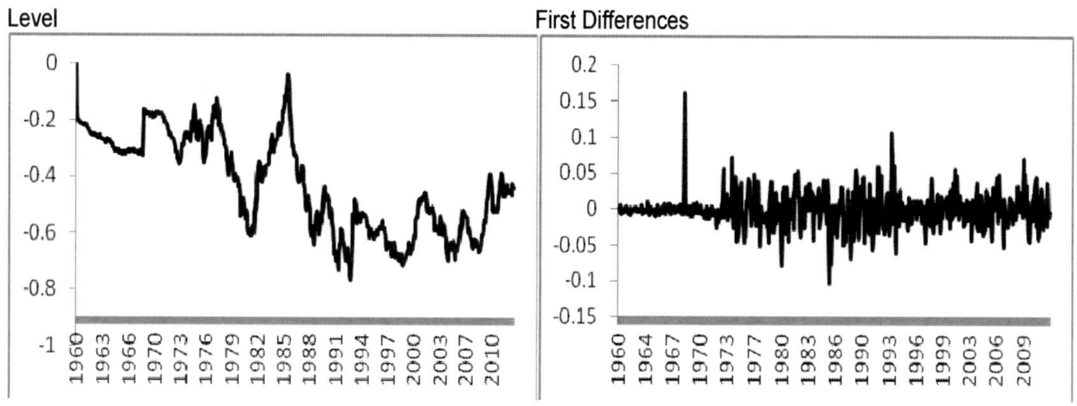

Figure 1. USD/GBP Real Exchange Rate (in Natural Logarithms).

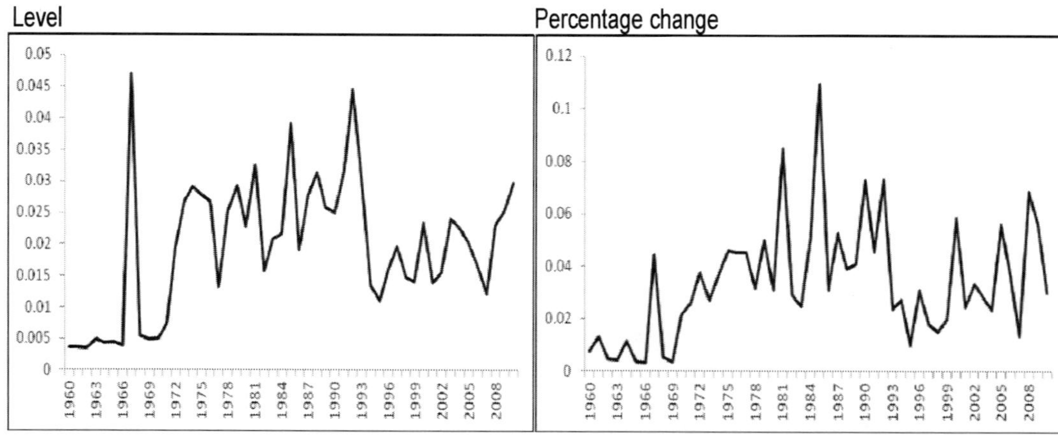

Figure 2. Annual Volatility Measures: Standard deviations of monthly observations.

We begin by grouping our constructed measures into the three groups that we outlined above: Variance-based measures, those based on prediction error, and measures created via GARCH models. Within each group, we also create a number of alternative measures by varying the window length or other appropriate parameters. We find that while each measure appears to capture similar periods of increased variability, the level of persistence after a shock differs from measure to measure. In addition, the degree of correlation among measures is highly inconsistent.

Since the log real rate used here is nonstationary (the corresponding Phillips-Perron statistic is -2.17), we use its first difference as the basis of almost all of our models. One exception is one of the annual volatility measures, which is used by Bahmani-Oskooee and Hegerty (2009a) and others in models that allow for nonstationary data. Since their cointegration methods permit the use of I(1) variables, they do not difference their exchange rates before creating a volatility series. We begin with these annual models, which are depicted in Figure 2.

In Figure 2, we see that the four main periods of increased volatility that we mentioned above are captured by both annual volatility measures—using both the level and the percentage changes in the exchange rate. The general shape is similar as well. One important difference between the two variations is that the overall level of volatility appears to be higher when levels data are used. The strong pound of the 1980s corresponds with a certain amount of extra volatility that is not present in the series created with the differenced data. This gives support to the idea that an exchange rate's variance is related to its level, and suggests that future researchers base such series on percentage changes in the appropriate series.

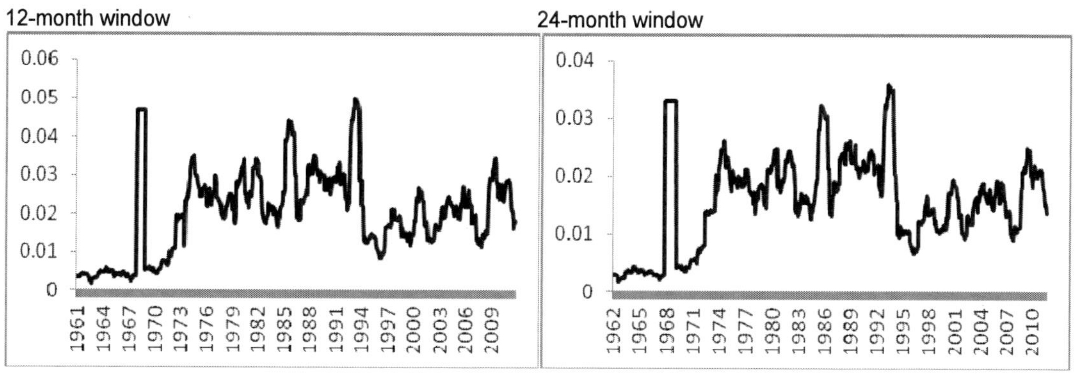

Figure 3. Monthly Volatility Measures Based on Moving Standard Deviations.

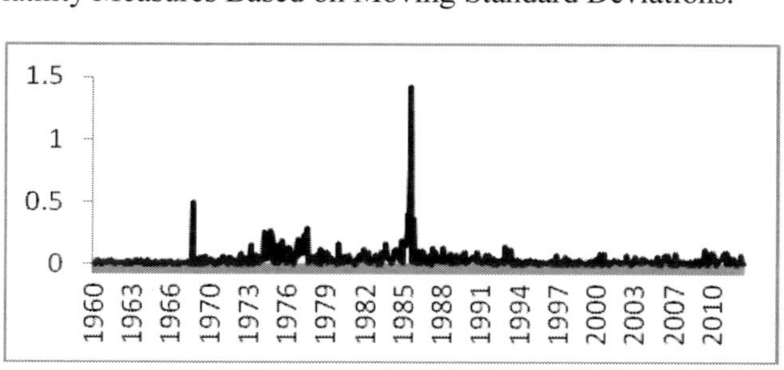

Figure 4. Monthly Volatility Measures Based on Absolute Value of Percentage Changes.

Table 1. AR(1) Estimation of Selected Volatility Series

	Measure						
Coefficient	*MASD12*	*MASD24*	*AR1_12*	*AR1_24*	*ARMARES*	*QTRES*	*APC*
	0.021	0.016	0.006	0.004	0.001	0.001	0.046
C	(0.000)	(0.000)	(0.000)	(0.000)	(0.000)	(0.000)	(0.000)
	0.952	0.958	0.881	0.949	0.098	0.076	0.463
AR(1)	(0.000)	(0.000)	(0.000)	(0.000)	(0.017)	(0.065)	(0.000)

p-values in parentheses.

Table 2. GARCH Coefficient Estimates

Specification GARCH(1,1)		GARCH-M(1,1)		EGARCH(1,1)		EGARCH-M(1,1)	
Coefficient	Estimate (p-value)	Coefficient	Estimate (p-value)		Coefficient	Estimate (p-value)	Coefficient
			-3.605				7.028
		GARCH	(0.493)			GARCH	(0.006)
	0.000		0.001		0.000		-0.003
C	(0.826)	C	(0.638)	C	(0.842)	C	(0.008)
	-0.999		-0.999		-0.971		-0.959
AR(1)	(0.000)	AR(1)	(0.000)	AR(1)	(0.000)	AR(1)	(0.000)
	1.271		1.273		1.273		1.240
MA(1)	(0.000)	MA(1)	(0.000)	MA(1)	(0.000)	MA(1)	(0.000)
	0.280		0.282		0.295		0.283
MA(2)	(0.000)	MA(2)	(0.000)	MA(2)	(0.000)	MA(2)	(0.000)
	0.000		0.000		-3.852		-0.083
α_0	(0.214)	α_0	(0.205)	α_0	(0.025)	α_0	(0.000)
	0.128		0.128		0.314		-0.041
α_1	(0.014)	α_1	(0.015)	α_1	(0.006)	α_1	(0.000)
	0.703		0.704		-0.040		-0.043
β_1	(0.000)	β_1	(0.000)	λ_1	(0.537)	λ_1	(0.159)
					0.521		0.985
				β_1	(0.027)	β_1	(0.000)
Akaike Info. Criterion	-4.832		-4.830		-4.825		-4.985

p-values in parentheses.

Figure 3 depicts the moving standard deviations (of percentage changes) in the pound-dollar real rate. We construct them using both 12- and 24-month windows. Both appear to be very similar upon visual inspection. As might be expected, the longer window results in wider volatility “spikes” as innovations persist longer before dropping out of the rolling estimation. Perhaps as a result, the magnitudes of these increases are reduced in the 24-month estimations. As shown in Table 1, both time series are highly autocorrelated, with the value on the variable with the longer window slightly higher. This finding suggests that time-series analyses that use this measure must take this potential autocorrelation into account; perhaps the use of robust standard errors (such as those introduced by Newey and West, 1987) can help control for such problems. The autocorrelation of the measure based on absolute percentage changes in the real rate, depicted in Figure 4, is somewhat lower, but seems to put a large emphasis on the mid-1980s dollar depreciation.

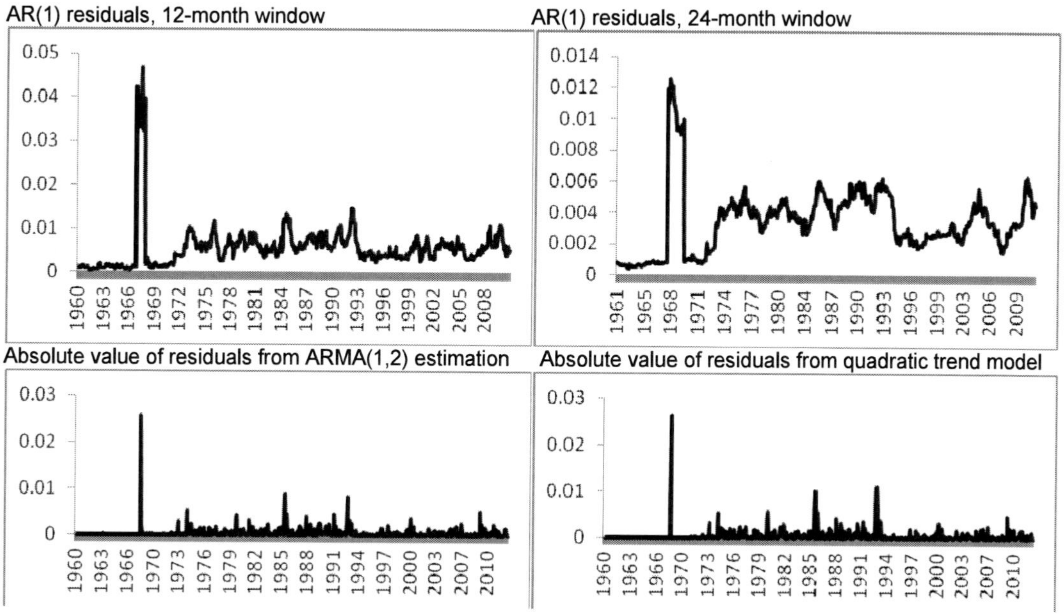

Figure 5. Monthly Volatility Measures Based on Prediction Error.

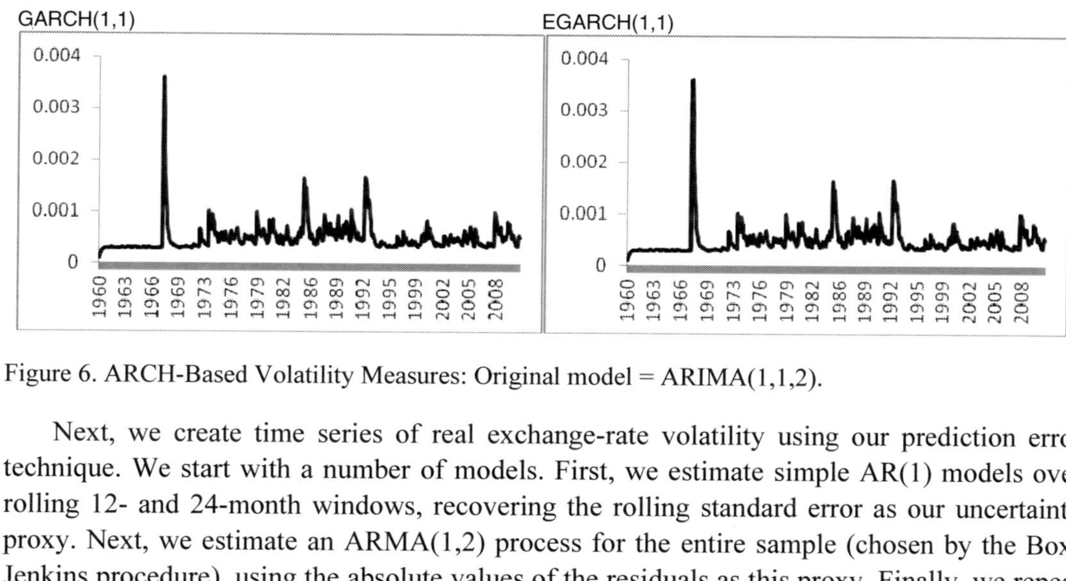

Figure 6. ARCH-Based Volatility Measures: Original model = ARIMA(1,1,2).

Next, we create time series of real exchange-rate volatility using our prediction error technique. We start with a number of models. First, we estimate simple AR(1) models over rolling 12- and 24-month windows, recovering the rolling standard error as our uncertainty proxy. Next, we estimate an ARMA(1,2) process for the entire sample (chosen by the Box-Jenkins procedure), using the absolute values of the residuals as this proxy. Finally, we repeat this process with the quadratic trend model of Equation (4) before plotting the absolute values of the residuals.

These four volatility proxies are presented in Figure 5. We see that those that are based on rolling windows show more persistence in the volatility series, with that of the 24-month window exceeding that of the 12-month window. The absolute values of the residuals, produced using our two full-sample models, do not show such carryovers. They do, however, show more intense "spikes" during those four periods discussed previously. Table 1 again shows us that the rolling estimations introduce a large degree of autocorrelation into the volatility proxies, which must be controlled for when these are used in time-series analyses.

Table 3. Correlations Among Volatility Measures

	GARCH	EGARCH	MASD12	MASD24	AR1_12	AR1_24	ARMARES	QTRES	APC
GARCH	1	0.861	0.670	0.630	0.614	0.526	0.098	0.086	0.106
EGARCH		1	0.514	0.492	0.393	0.363	0.109	0.091	0.113
Std. Dev. (12-month)			1	0.960	0.680	0.664	0.088	0.096	0.147
Std. Dev. (24-month)				1	0.637	0.695	0.082	0.088	0.135
AR(1) Standard Errors (12)					1	0.725	0.309	0.281	0.165
AR(1) Standard Errors (24)						1	0.236	0.225	0.156
ARIMA Residuals							1	0.970	0.562
Quadratic Trend Residuals								1	0.595
Absolute Percentage Changes									1

Finally, we construct volatility measures for certain GARCH variations as follows. First, we our ARIMA(*1,1,2*) model for monthly values of the pound-dollar real exchange rate as in Equation (5). We then test for ARCH errors using the Lagrange Multiplier test of Engle (1982). Since we find evidence of volatility clustering, we estimate GARCH(1,1) and EGARCH(1,1) processes for the error terms. We also examine GARCH-in-Mean for both, but only create volatility series if the volatility term is significant in the mean equation.

The GARCH coefficients are provided in Table 2. We see that for the standard GARCH(1,1) there is no evidence of any GARCH-in-Mean effect. Uncertainty does not seem to have an impact on the real exchange rate itself. At the same time, Exponential GARCH does have GARCH-M effects. But because of an insignificant coefficient and after visually inspecting the variance series, we omit this result. Instead, we focus on GARCH(1,1) and EGARCH(1,1). Most likely, the former is preferred, based on the significance of the coefficient estimates. We provide graphs of both resulting volatility series in Figure 6.

While these two series both capture the main periods of volatility, they are nearly identical visually. Just how similar are they? This presents the opportunity to begin to compare the common properties of these two series, as well as of all the others calculated in this study. We find in Table 3 that GARCH and EGARCH are highly correlated (their coefficient is 0.861). Even higher is the correlation between the two moving-standard-deviation measures (it is nearly 1.00). This suggests that window length is not that important in constructing this measure. Relatively uncorrelated are the two measures that capture prediction error using residuals: While nearly perfectly correlated with each other, these two regression-based models (ARIMA and trend) have only weak relationships with the other measures. Correlations are stronger for the measures that are based on rolling standard errors.

Overall, our analysis of these different volatility measures allows us to draw three main conclusions. First, each main approach—variance-based, prediction-error, and GARCH—is capable of capturing the same broad periods of increased uncertainty. In out example, these include the breakdown of Bretton Woods, the 1967 devaluation, and the 1992 ERM crisis. Second, while these measures capture similar events, they are not all closely associated with one another. Bilateral correlation coefficients tend to be higher for members of each sub-group. Finally, while moving standard deviations are popular in the literature on many types of macroeconomic volatility, they seem to be most prone to autocorrelation.

CONCLUSION

Of the many possible ways to measure "uncertainty" in economics, no single measure has come to dominate the literature. Yet there are relatively few systematic analysis that compare alternative measures side-by-side. In this chapter, we focus on one narrow classification of this concept: Those that are used in empirical, time-series, macroeconomic studies. Even more narrowly, we concentrate on exchange-rate volatility.

Drawing on out 2007 literature review, we break these measures down into three "families." First, we examine those that calculate time series' volatility, usually through standard deviations. Next, we look at measures that might more truly calculate "uncertainty" by quantifying the prediction error of an underlying model. Finally, we study ARCH models.

After discussing each measure in turn, we apply these models to an analysis of the monthly dollar-pound real exchange rate from 1960 to 2011.

We find that while almost all these measures capture similar periods of high and low volatility, especially those that match actual events in the foreign exchange markets. We also find that certain variations of each method within each family, such as varying a window length or using a different GARCH technique, lead to only small changes in the calculated volatility series. But certain characteristics differ greatly between groups. In particular, the degree of persistence (measured as a first-order autoregressive process) is greater for models that use moving standard deviations, and prediction-error models lead to series that are not correlated with competing measures. This has important implications for practitioners who wish to correctly specify models that require the choice of a proxy for uncertainty.

REFERENCES

Ahmed, Abdullah D. and Sandy Suardi (2009), "Macroeconomic Volatility, Trade and Financial Liberalization in Africa," *World Development* 37(10), 1623-1636.

Akhtar, M.A. and R.S. Hilton (1984), "Effects of Uncertainty on German and U.S. Trade," *Federal Reserve Bank of New York Quarterly Review*, Spring 1984, 7-16.

Arize, A. and D. K. Ghosh (1994), "Exchange-Rate Uncertainty and Recent U.S. Export Demand Instability," *The International Trade Journal* 7(3), 347-365.

Asseery, A., Peel, D.A. (1991), "The Effects of Exchange Rate Volatility on Exports—Some New Estimates," *Economics Letters* 37, 173–177.

Bahmani-Oskooee, Mohsen and Scott W. Hegerty (2007), "Exchange Rate Volatility and Trade Flows: A Review Article," *Journal of Economic Studies* 34(3), 211-255.

Bahmani-Oskooee, M., and S. W. Hegerty (2009a), "The Effects of Exchange-Rate Volatility on Commodity Trade Between the U.S. and Mexico," *Southern Economic Journal* 79, 1019-1044.

Bahmani-Oskooee, M., and S. W. Hegerty (2009b), "Exchange-Rate Risk and U.S.-Japan Trade: Evidence from Industry Level Data," *Journal of the Japanese and International Economies* 22(4), 518-534.

Bahmani-Oskooee, Z. Ardalani, and M. Bolhassani (2010), "Exchange-Rate Volatility and U.S. Commodity Trade with the Rest of the World," International Review of Applied Economics 24, 511-532.

Bahmani-Oskooee, M., M. Bolhasani, and S.W. Hegerty (2011), "Exchange-Rate Volatility and Industry Trade Between Canada and Mexico," *The Journal of International Trade & Economic Development*, in press.

Bailey, M.J., Tavlas, G.S., and M. Ulan, (1987), "The Impact of Exchange Rate Volatility on Export Growth: Some Theoretical Considerations and Empirical Results", *Journal of Policy Modeling*, 9, 225-243.

Bollerslev, T. (1986), "Generalized Autoregressive Conditional Heteroscedasticity, *Journal of Econometrics* 31, 307-327.

Broner, F.A. and R. Rigobon (2005), "Why are Capital Flows So Much More Volatile in Emerging Than in Developed Countries?," Central Bank of Chile Working Paper #328.

Chowdhury, Abdur R, 1993. "Does Exchange Rate Volatility Depress Trade Flows? Evidence from Error-Correction Models," The Review of Economics and Statistics, MIT Press, vol. 75(4), pages 700-706.

De Grauwe, P. (1988), "Exchange Rate Variability and the Slowdown in Growth of International Trade." *International Monetary Fund Staff Papers* 35, 63–84.

Edwards, S. (1998), "Interest Rate Volatility, Contagion and Convergence: An Empirical Investigation of the Cases of Argentina, Chile and Mexico," *Journal of Applied Economics* 1(1), 55-86.

Edwards, S. and R. Susmel (2003), "Interest-Rate Volatility in Emerging Markets," *The Review of Economics and Statistics* 85(2), 328-348.

Engle, R.F. (1982) "Autoregressive Conditional Heteroscedasticity With Estimates of the Variance of the United Kingdom Inflation," *Econometrica* 50, 377-403.

Engle, R.F., D.M. Lilien, and R.P. Robins (1987), "Estimating Time Varying Risk Premia in the Term Structure: The ARCH-M Model," *Econometrica* 55, 391-407.

Fang, Wen-Shwo and Stephen M. Miller, (2008), "The Great Moderation and the Relationship between Output Growth and Its Volatility," *Southern Economic Journal* 74(3), 819-838.

Grobar, L.M. (1993), "The Effect of Real Exchange Rate Uncertainty on LDC Manufactured Exports," *Journal of Development Economics* 41(2), 367-376.

Kenen, Peter B. and Dani Rodrik (1986), "Measuring and Analyzing the Effects of Short-Term Volatility in Real Exchange Rates," *The Review of Economics and Statistics* 68, 311-315.

Laopodis, N.T. (2000), "Monetary Policy Implications of Volatility Linkages Among Long-Term Interest Rates," *Journal of Economics and Finance* 24(2), 160-177.

Laopodis, N.T. (2001), "International Interest-Rate Transmission and the 'German Dominance Hypothesis' Within EMS," *Open Economies Review* 2(4), 347-377.

Laopodis, N.T. (2002), "Volatility Linkages Among Interest Rates: Implications for Global Monetary Policy," *International Journal of Finance and Economics* 7(3), 215-233.

Nelson, D.B. (1991) "Conditional Heteroskedasticity in Asset Returns: A New Approach," *Econometrica* 59(2), 347-370.

Newey, W.K., and K.D. West (1987), "A Simple, Positive Semi-Definite, Heteroskedasticity and Autocorrelation Consistent Covariance Matrix," *Econometrica* 55(3), 703–708.

Ramey, G. and V. A. Ramey (1995), "Cross-Country Evidence on the Link Between Volatility and Growth," *American Economic Review* 85(5), 1138-1151.

Sauer, C. and A.K. Bohara, (2001), "Exchange Rate Volatility and Exports: Regional Differences Between Developing and Industrialized Countries," *Review of International Economics* 9(1), 133-152.

Thursby, J.G. and Thursby, M. (1985), "The Uncertainty Effects of Floating Exchange Rates: Empirical Evidence on International Trade Flows," in Arndt, S.W., Sweeney, R.J. and Willett, T.D. (Eds), *Exchange Rates, Trade and the US Economy*, Ballinger, Cambridge, MA.

In: Economics of Innovation, Incentives and Uncertainty ISBN: 978-1-62257-251-9
Editors: P. Simmons and S. T. Jordan

Chapter 8

STRATEGIC MANAGERIAL DELEGATION AND ENVIRONMENTAL TAXES IN MIXED OLIGOPOLY

Leonard F. S. Wang* and Ya-Chin Wang
[1]Department of Applied Economics, National University of Kaohsiung, Kaohsiung, Taiwan
[2]Department of Finance and Banking, Kun Shan University, Tainan, Taiwan

ABSTRACT

This paper examines the implications of strategic delegation in a mixed oligopoly model when the government imposes an environmental tax to mitigate pollution damage. Managers of both public and private firms are compensated by a linear combination of profit and sales revenue. The effects of strategic delegation by comparing the equilibria with and without strategic delegation are explored. It shows that when strategic delegation is applied, the government sets a lower environmental tax. It is opposite to Bárcena-Ruiz and Garzón's (2002) results in pure oligopoly with managerial delegation, but concurs with the finding obtained in Bárcena-Ruiz and Garzón (2006) and Wang and Wang (2009) in mixed market settings without managerial delegation.

Keywords: Environmental Taxes, Strategic Delegation, Mixed Oligopoly

1. INTRODUCTION

Since the 1990s, concerns over environmental quality have been prominent in the public domain. Government conducts environmental policies by imposing taxes and/or direct regulation on adoption of costly abatement technology by the firms for mitigating pollution

* Tel: 886-7-5919322, Fax:886-7-5919320, E-mail: lfswang@nuk.edu.tw.

damage. In many transition economies, state-owned firms often play a significant role in mixed markets existing in a broad range of industries such as oil, telecommunications, steel, electricity, airline and railway, which were formerly dominated by public monopolies. As a result, mixed oligopoly models have received considerable attention in economic research.[1]

In the existing literature of environmental issues under oligopoly, Simpson (1995) derived the optimal pollution for a Cournot duopoly and found that if firms have different production costs, the optimal tax rate may exceed the marginal damage. Bárcena-Ruiz and Garzón (2002) analyzed the environmental policy in a pure duopolistic model with incentive schemes showing that firm owners have to pay a higher environmental tax and both environmental damage and social welfare increase compared to the profit-maximization case. Bárcena-Ruiz and Garzón (2006) demonstrated that the outcome of the decision whether to privatize a public firm may be different if the government internalizes the environmental damage than if the government ignores it. When the government sets a tax to protect the environment, the tax is lower in the mixed oligopoly than in the private one even though the environmental damage is greater. Wang and Wang (2009) considered the effect of privatization on environmental damage and social welfare in a differentiated mixed duopolistic framework. They found that the environmental tax is lower in the mixed duopoly than in the private one, while Wang *et al.* (2009a) in a partially privatized mixed duopoly demonstrated that privatization unambiguously reduces the pollution levels of firms.

In a different strand of the literature, the theory of strategic managerial delegation started with Vickers (1985), Fershtman and Judd (1987) and Sklivas (1987) (henceforth VFJS). It is well known that the separation of ownership and management provides firms with the opportunity to commit to non-profit maximizing decisions leading to more aggressive managerial behavior than in the standard Cournot case.[2] Barros (1995) extended strategic delegation to the field of mixed oligopoly, and argued that "since social welfare is a very complex concept, which depends also upon rivals' actions, it seems unreasonable to suppose that the public agent is offered such a contract" (P. 374).

He examined a mixed duopolistic model in which both public and private firms base the manager's compensation upon a linear combination of profit and sales revenue. Saha and Sensarma (2008) studied a mixed oligopoly where a partially public firm competes with a private firm, and found that when the private firm offers managerial incentives, there is a redistribution of profit and output from the private to the public firm, but the aggregate output and social welfare may remain unchanged. When the private firm is owned overseas, the extent of privatization is less and managerial incentives are milder. Heywood and Ye (2009) recently derived the optimal incentive contract for a public firm that weighs both profit and

---

[1] This is evidenced by the applications of these models to partial privatization by Matsumura (1998), to endogenous order of moves by Pal (1998), and to the location choice by Matsushima and Matsumura (2006). See, for example, Goering (2008) recently analyzed three separate Cournot mixed market scenarios: competition between the non-profit firm (NPO) and a private for-profit firm; competition between the NPO and a public firm; and a market scenario that includes all three firms.

[2] See, for example, Macho-Stadler and Verdier (1991), Hwang and Mai (1995), and Das (1997). The separation of ownership and management may also exist in a public firm taking up this challenge. Another way to take up such a great challenge is to privatize the public firm, which improves its productivity. De Fraja and Delbono (1989) showed that a public firm should be privatized in some cases to improve social welfare. Privatization is the transfer of shareholdings while strategic delegation is the transfer of management rights.

welfare. They showed that its use may either increase or decrease welfare depending on the number of private firms and the exact nature of costs.[3]

In this paper, we examine environmental taxation in a mixed oligopolistic model in which a public firm competes with one or more private firms. The government sets an environmental tax in advance. Taking the government's policy as announced, owners of the firms then compete in the market by designing incentive schemes.

As it is specified in most of the literature, the objective of the public firm owner is social welfare maximization while that of the private one is purely profit maximization. In the end, managers of firms engage in competition by making output and abatement decisions. Following Barros (1995), Heywood and Ye (2009), and Bárcena-Ruiz (2010), the managers of both public and private firms are compensated by a linear combination of profit and sales revenue. This paper derives the optimal environmental tax and the equilibrium values of other variables with and without strategic delegation to show the effects of strategic delegation. It demonstrates that when strategic delegation is applied, the government sets a lower environmental tax. This result is opposite to Bárcena-Ruiz and Garzón (2002), but concurs with the findings of Bárcena-Ruiz and Garzón (2006), and Wang and Wang (2009).

The remainder of this paper is organized as follows. Section 2 presents the model under strategic delegation. Section 3 seperately deals with the no-delegation game as a benchmark and the game with incentive contract. Section 4 analyzes the effects of strategic delegation through a comparison of equilibrium outcomes. Concluding remarks are provided in Section 5.

2. The Model

2.1. Basic Setting

We depict a market made up of one public firm (indexed by 0) and n private firms (indexed by $i, i = 1, ..., n$) producing a homogeneous good. We keep Ulph's (1996) assumptions, providing that each unit of the good produced, given by q_j, leads to one unit of pollution e_j, $j = 0, ..., n$. The environmental damage is measured by a quadratic form: $ED = (\Sigma_j e_j)^2 / 2$. However, each firm can prevent pollution by undertaking abatement measures. If firm j chooses pollution abatement level a_j, then the emission level of each firm is $e_j = q_j - a_j$. It is reasonable to assume that pollution emission of each firm is non-negative. The cost of pollution abatement of firm j is $a_j^2 / 2$. Each firm has to pay an environmental tax t per unit of pollutant emitted and as a result, the total taxes obtained by the government are $T = t \Sigma_j e_j$.

[3] In different context, Tomaru *et al.* (2009) showed that optimal subsidies attain the first-best allocation in Cournot mixed and private duopolies and in mixed duopolies wherein the public firm acts as the leader with respect to the decision of the quantity level or the weight of each firm's output in the delegation contract. Tomaru and Saito (2009) observed that including the production subsidy and endogenous timing in mixed oligopoly, private leadership may disappear from equilibrium and that Cournot and public leadership become the likely equilibrium outcomes.

We define the total supply from private firms as $Q_p = \Sigma_i q_i$ .The price of this commodity is characterized by a linear inverse demand function $p = A - Q$, where $Q = q_0 + Q_p$. Accordingly, total consumer surplus is $CS = Q^2/2$. For simplicity, we assume that direct costs of production of all firms are absent. Therefore, the profit functions of the public firm and each private firm are

$$\pi_0 = pq_0 - te_0 - a_0^2/2 \tag{1}$$

and

$$\pi_i = pq_i - te_i - a_i^2/2, i = 1,....,n. \tag{2}$$

2.2. Strategic Incentive Contract

Each owner hires a manager offering the contract that is a linear combination of the firm's gross profit and sales revenue. The manager is authorized to make decisions on production and pollution abatement. Using the framework of VFJS, managers are risk neutral and maximize the following compensation function:

$$R_0 = \alpha_0 \pi_0 + (1 - \alpha_0) pq_0 \tag{3}$$

and

$$R_i = \alpha_i \pi_i + (1 - \alpha_i) pq_i, \; i = 1,....,n. \tag{4}$$

where α_0 and α_i's are the incentive parameters designed by their owners.[4] For values of $\alpha < 1$ ($\alpha > 1$) manager receives an incentive to be more aggressive (less aggressive) than a profit-maximizer. No range limitation is imposed on the values of α as long as each firm produces positive output.

2.3. Mixed Oligopoly

As in Barros (1995) and White (2001), the public firm here is a typical "public-owned and private-operated" enterprise. Public ownership requires that the firm pursue goals other than profit maximization, but ownership alteration does not imply any change in private operation in market stage. On the other hand, the objective function of the public firm, i.e.

[4] The objective function we've used here mainly takes from Barros' (1995, Eq. (5)) model on mixed incentive game. White (2001), Bárcena-Ruiz (2009, 2010) and Tamaru *et al.* (2009) also considered the managerial incentive contract when public and private firms compete in the mixed oligopoly market.

social welfare, is specified for the public owner to use as a regulation mechanism when it delegates private operator competing in the market.[5]

It might be asked that what the major role of the government is playing. The environmental tax is imposed by the government and its goal is to maximize social welfare as the public owner does. We adopt Bárcena-Ruiz and Garzón's (2002, 2006) definition, the welfare function comprises the consumer surplus, CS, the producer surplus, $\pi_0 + \sum_i \pi_i$, the total taxes collected by the government T, and the environmental damage caused by the production process, ED:

$$W = CS + \pi_0 + \sum_i \pi_i + T - ED. \tag{5}$$

3. The Scenarios and Equilibrium Outcomes

3.1. Benchmark Case: Mixed Oligopoly without Managerial Incentive

A no-delegation game, which consists of two stages, is analyzed here first as a benchmark. In the first stage, the government sets the environmental tax. In the second stage, each firm chooses the output and pollution abatement levels. The game is solved by backward induction. We denote the equilibrium of mixed oligopoly without managerial incentive by the superscript *MN*.

In the second stage, the managers of public and private firms make the output and pollution abatement decisions to maximize Eqs.(5) and (2), respectively. The solutions of this optimization problem are given by:

$$a_0 = \frac{A - nt}{3},\ a_i \equiv a = t,\ q_0 = \frac{A(2-n) + (4n + n^2)t}{3},\ q_i \equiv q = \frac{A - (3+n)t}{3}.$$

In the first stage, using the results in the second stage, the government sets the environmental tax to maximize Eq. (5). Solving the first-order condition and given the satisfaction of second-order condition gives the unrestricted optimal environmental tax:

$$t = \frac{A}{3+n}.$$

However, the environmental tax set by the government can be neither too high nor too low for the restriction of non-negative emission level of both public and private firms. That is

$$t_0 \equiv \max\{\frac{A(n-1)}{n(n+5)}, 0\} \le t \le \frac{A}{6+n} \equiv t_1.$$

[5] The same objective function can be seen in Bárcena-Ruiz and Garzón's (2002, Eq.(3)) with environmental tax consideration and Bárcena-Ruiz's (2006, Eq. (2)) under environmental regulation.

Therefore, the optimal environmental tax under the restriction of the non-negative emission level is

$$t^{MN} = t^* = \frac{A}{6+n}.$$

Hence, we obtain the following lemma.

Lemma 1. In a mixed oligopoly without managerial incentive, in equilibrium, the optimal environmental tax, the abatement level of each firm, the output level of each firm, the total output, the market price, the profit of each firm, the environmental damage and the social welfare are, respectively:

$$t^{MN} = \frac{A}{6+n},\quad a^{MN} = \frac{A}{6+n},\quad a_0^{MN} = \frac{2A}{6+n},\quad q^{MN} = \frac{A}{6+n},\quad q_0^{MN} = \frac{4A}{6+n},$$

$$Q^{MN} = \frac{A(4+n)}{6+n},\quad p^{MN} = \frac{2A}{6+n},\quad \pi^{MN} = \frac{3A^2}{2(6+n)^2},\quad \pi_0^{MN} = \frac{9A^2}{2(6+n)^2},$$

$$ED^{MN} = \frac{2A^2}{(6+n)^2},\quad W^{MN} = \frac{A^2(22+11n+n^2)}{2(6+n)^2}.$$

3.2. Mixed Oligopoly with Incentive Contract

The incentive game consisting of three stages is analyzed in this subsection. In the first stage, the government sets the environmental tax. In the second stage, the owners compete in designing managerial incentives. In the third stage, the managers compete in the market by choosing output and abatement levels. We denote the equilibrium of mixed oligopoly with managerial incentive by the superscript *MI*. The game is also solved by backward induction. In the third stage, the managers compete in the market by choosing output and abatement levels. From the first-order conditions and considering the symmetric assumption, we obtain that:

$$\frac{\partial R_0}{\partial q_0} = A - 2q_0 - \Sigma_i q_i - \alpha_0 t = 0 \quad \text{and} \quad \frac{\partial R_i}{\partial q_i} = A - q_0 - \Sigma_i q_i - q_i - \alpha_i t = 0.$$

$$\frac{\partial R_0}{\partial a_0} = \alpha_0 (t - a_0) = 0 \quad \text{and} \quad \frac{\partial R_i}{\partial a_i} = \alpha_i (t - a_i) = 0.$$

Thus, the output levels and the abatement levels here are functions of the incentive schemes designed in the previous stage:[6]

---

6 Please note that we do not introduce the symmetric equilibrium assumption here. Therefore, the output of a representative private firm and total private output are both required in this stage.

$$q_0 = \frac{A + t\Sigma_i \alpha_i - (n+1)t\alpha_0}{n+2}, \; Q_p = \frac{nA + nt\alpha_0 - 2t\Sigma_i \alpha_i}{n+2},$$
$$q_i = \frac{A + t\Sigma_i \alpha_i + t\alpha_0}{n+2} - \alpha_i t, \; a_0 = a_i = t. \tag{6}$$

In the second stage, the owners choose optimal incentive schemes for the managers. By maximizing Eq. (5) with respect to α_0 and Eq. (2) with respect to α_i and confining our attention to symmetric equilibrium[7], we obtain the optimal incentive schemes:

$$\alpha_0(t) = \frac{(1+n)[nA - (2+3n+n^2)t]}{2t}, \; \alpha(t) = \frac{(2+3n+n^2)t - nA}{2t}. \tag{7}$$

In the first stage, the government sets the environmental tax to maximize social welfare. Using the results obtained, we can express the outputs and welfare in terms of t:

$$q_0(t) = \frac{1}{2}[(1+4n+4n^2+n^3)t + A(1-n-n^2)],$$
$$q(t) = \frac{1}{2}(1+n)[A - t(3+n)]. \tag{8}$$

As in the section 3.1, solving $\partial W(t)/\partial t = 0$ and given that $\partial^2 W(t)/\partial t^2 < 0$, the unrestricted optimal environmental tax is found as:

$$\hat{t} = \frac{A}{3+n}.$$

Likewise, the environmental tax the government sets must satisfy the condition of non-negative emission level:

$$\hat{t}_0 \equiv \frac{A(-1+n+n^2)}{-1+4n+4n^2+n^3} \le \hat{t} \le \frac{A(1+n)}{5+4n+n^2} \equiv \hat{t}_1.$$

Therefore, the optimal environmental tax is

$$t^{MI} = \hat{t}^* = \frac{A(1+n)}{5+4n+n^2}.$$

After substitution, the following lemma is obtained.

Lemma 2. In a mixed oligopoly with managerial incentives and environmental policy, in equilibrium, the optimal environmental tax, the incentive scheme of each firm, the abatement

[7] We confine our attention to symmetric equilibrium, i.e., $\Sigma_i \alpha_i = n\alpha$. Then according to Eq.(6), $Q_p = \Sigma_i q_i = nq$. Therefore, the subscript i disappears in what follows.

level of each firm, the output level of each firm, the total output, the market price, the profit of each firm, the environmental damage and the social welfare are, respectively:

$$t^{MI} = \frac{A(1+n)}{5+4n+n^2}, \ \alpha_0^{MI} = -1, \quad \alpha^{MI} = \frac{1}{1+n}, \quad a_0^{MI} = a^{MI} = \frac{A(1+n)}{5+4n+n^2},$$

$$q^{MI} = \frac{A(1+n)}{5+4n+n^2}, \quad q_0^{MI} = \frac{A(3+2n)}{5+4n+n^2}, \quad Q^{MI} = \frac{A(3+3n+n^2)}{5+4n+n^2}, \quad p^{MI} = \frac{A(2+n)}{5+4n+n^2},$$

$$\pi^{MI} = \frac{A^2(3+4n+n^2)}{2(5+4n+n^2)^2}, \ \pi_0^{MI} = \frac{A^2(7+6n+n^2)}{2(5+4n+n^2)^2},$$

$$ED^{MI} = \frac{A^2(2+n)^2}{2(5+4n+n^2)^2}, \ W^{MI} = \frac{A^2(16+29n+21n^2+7n^3+n^4)}{2(5+4n+n^2)^2}.$$

In Lemma 2, negative α_0^{MI} implies that the manager of the public firm is penalized for higher pollution brought by over producing. The over-production of the public firm is owing to its aim of pursuing social welfare maximization. This result is also pointed out by Barros (1995) in a mixed duopoly market. Chang (2007) also argued that in an international mixed duopoly the public firm may earn negative profit.

4. Comparison of the Equilibrium Outcomes

4.1. Effects of Incentive Contract

In the previous sections, we have derived equilibria in mixed oligopoly with and without managerial incentive. By comparing Lemma 2 with Lemma 1 (the benchmark case), we obtain the effects of strategic delegation.

Proposition 1. In a mixed oligopoly with environmental policy:

(1) $q_0^{MN} > q_0^{MI} > q^{MI} > q^{MN}$, $Q^{MN} < (>) Q^{MI}$ and $p^{MN} > (<) p^{MI}$, if $n \geq (<) 2$;

(2) $t^{MN} > t^{MI}$, $a_0^{MN} > a_0^{MI}$, $a^{MN} < a^{MI}$ and $ED^{MN} > (<) ED^{MI}$, if $n \geq (<) 2$;

(3) $\pi_0^{MN} > \pi_0^{MI}$ and $\pi^{MN} > (<) \pi^{MI}$ if $n \geq (<) 3$;

(4) $W^{MN} < W^{MI}$.

Due to an obligation towards social welfare, the public firm tends to produce more than private firms do in the market, no matter whether strategic delegation is applied or not. Given that each owner directs his manager to more aggressive behavior by using nonprofit incentives, each private firm produces more than does the no-delegation game. However, the manager is forced to reduce her output because of all competing managers being more aggressive, $q_0^{MN} > q_0^{MI} > q^{MI} > q^{MN}$. When $n = 1$, the public firm with higher output in the no-delegation game is a dominant one in the whole market; nonetheless, when $n \geq 2$, it is the private one with higher output in the delegation game that dominates. Therefore,

$Q^{MN} < (>) Q^{MI}$ and $p^{MN} > (<) p^{MI}$, if $n \geq (<) 2$. Accordingly, a decrease of public output in a comparison with the no-delegation game lowers public profit, $\pi_0^{MN} > \pi_0^{MI}$. In our model, when $n = 1$ and $n = 2$, $\pi^{MN} < \pi^{MI}$. Nevertheless, the increase of private output in delegation game may highly possible make each private firm gain lower profit in equilibrium ($\pi^{MN} > \pi^{MI}$ if $n \geq 3$). As is well known, strategic delegation is a profit-shifting mechanism made by the owners, which reduces the scale of government's intervention.[8] The result we have is that $t^{MN} > t^{MI}$ which confirms this well-known statement, and in the delegation game, the public firm has a lower abatement level ($a_0^{MN} > a_0^{MI}$). However, strategic delegation triggers more private outputs, which in turn requires a higher abatement level ($a^{MN} < a^{MI}$). The total emission level depends negatively and mainly on the total output, $ED^{MN} > (<) ED^{MI}$ if $n \geq (<) 2$.

In a mixed duopolistic model ($n = 1$) with managerial incentives, the welfare gain from higher private profit can outweigh the welfare loss from other considerations. In a mixed oligopolistic model ($n \geq 2$) with managerial incentives, the consumer surplus is higher while the environmental damage is lower, and altogether offset the lower producer surplus. Consequently, strategic delegation is beneficial for social welfare ($W^{MN} < W^{MI}$).

4.2. A Duopoly Case

The next question that we address is the following: Is the incentive contract in our mixed model socially preferable to a private incentive mechanism?

Table 1. Mixed and private duopoly

Mixed Duopoly (n=1)	Cournot Duopoly (Bárcena-Ruiz and Garzón, 2002)
$q_0^{MND} > q_0^{MID} > q^{MID} > q^{MND}$,	$q^{CID} > q^{CND}$,
$Q^{MID} < Q^{MND}$, $p^{MID} > p^{MND}$,	$Q^{CID} > Q^{CND}$, $p^{CID} < p^{CND}$,
$\pi_0^{MID} < \pi_0^{MND}$, $\pi^{MID} > \pi^{MND}$,	$\pi^{CID} < \pi^{CND}$,
$t^{MID} < t^{MND}$,	$t^{CID} > t^{CND}$,
$CS^{MID} > CS^{MND}$,	$CS^{CID} > CS^{CND}$,
$ED^{MID} > ED^{MND}$,	$ED^{CID} > ED^{CND}$,
$W^{MID} > W^{MND}$.	$W^{CID} > W^{CND}$.

Note: The superscripts represent the mixed duopoly without managerial incentive (*MND*), mixed duopoly with managerial incentive (*MID*), Cournot duopoly without managerial incentive (*CND*) and Cournot duopoly with managerial incentive (*CID*) respectively.

[8] See, for example, Das (1997), and Wang *et al.* (2008, 2009b).

In order to catch the spirit of the concerned issue and compare our results with the outcomes in Bárcena-Ruiz and Garzón (2002) which is the case of the private duopoly with managerial incentive, we work out the comparisons in Table 1.

As presented in Table 1, the incentive effects in mixed duopolistic model indicate a decrease in total output, environmental tax and public firm's profit, and an increase of private profit in comparison with that in the no-delegation game. These results are different from those in a pure Cournot duopoly reported by Bárcena-Ruiz and Garzón (2002): firm owners have to pay a higher environmental tax, while compared to the profit-maximization case both environmental damage and social welfare are increase. We would like to point out here that VFJS model shows that sale-revenue incentive is a dominant strategy, but the outcome is a "prisoner's dilemma" choice in the sense that both firms' profit are going down. In our mixed duopoly model, the private firm' profit rises when both firm implement the sale-revenue incentive contract, $\pi^{MID} > \pi^{MND}$. In addition, the findings of $Q^{MID} < Q^{MND}$ and $ED^{MID} > ED^{MND}$ are inasmuch as private managerial firm is so motivated by the incentive contract that it produces a lot. However, that is not enough to countervail the public managerial firm's less abatement and reduced production, so the total output decrease while the environment is damaged more; but it improves consumer surplus and social welfare, $CS^{MID} > CS^{MND}$ and $W^{MID} > W^{MND}$.

Conclusion

This paper elaborates the models of Bárcena-Ruiz and Garzón (2002, 2006). The motivation of this paper is to further investigate the optimal environmental policy and strategic delegation in a mixed oligopoly model by incorporating profit-shifting mechanism. The analyses yield some results that differ from those obtained in a pure duopoly and a mixed oligopoly.

The equilibrium outcomes derived with and without strategic delegation in a mixed oligopoly model, show that when strategic delegation is applied, the government sets a lower environmental tax. It is opposite to Bárcena-Ruiz and Garzón (2002) in pure oligopoly with managerial delegation, but concurs with the findings obtained in Bárcena-Ruiz and Garzón (2006) and Wang and Wang (2009) in mixed market settings without managerial delegation. The lower environmental tax impacts on both public's and private firms'decisions on output, pollution and abatement levels. It prevents the public firm from reducing its output level further, and makes private firms increase their output and abatement levels. Furthermore, strategic delegation corresponds with social welfare improvement.

References

Bárcena-Ruiz, J. C., 2009. The decision to hire managers in mixed markets under Bertrand competition. *Japanese Economic Review* 60, 376-388.

Bárcena-Ruiz, J. C., 2010. Strategic delegation and semipublic firms. *Economics Bulletin* 30 (1), 744-750.

Bárcena-Ruiz, J. C., Garzón, M. B., 2002. Environmental taxes and strategic delegation. *Spanish Economic Review* 4, 301-310.

Bárcena-Ruiz, J. C., Garzón, M. B., 2006. Mixed oligopoly and environmental policy. *Spanish Economic Review* 8, 139-160.

Barros, F., 1995. Incentive schemes as strategic variables: an application to a mixed duopoly. *International Journal of Industrial Organization* 13, 373-386.

Chang, W. W., 2007. Optimal trade, industrial, and privatization policies in a mixed duopoly with strategic managerial incentives. *Journal of International Trade and Economic Development* 16, 31-52.

Das, S. P., 1997. Strategic managerial delegation and trade policy. *Journal of International Economics* 43, 173-188.

De Fraja, G., Delbono, F., 1989. Alternatives strategies of a public enterprise in oligopoly. *Oxford Economic Papers* 41, 302-311.

Fershtman, C., Judd, K. L., 1987. Equilibrium incentives in oligopoly. *American Economic Review* 77, 927-940.

Goering, G. E., 2008. Welfare impacts of a non-profit firm in mixed commercial markets. *Economic Systems* 32, 326-334.

Heywood, J.S., Ye, G., 2009. Delegation in a mixed oligopoly: the case of multiple private firms. *Managerial and Decision Economics* 30, 71-82.

Hwang, H., Mai, C. C., 1995. Strategic management under duopoly. *Managerial and Decision Economics* 16, 239-247.

Macho-Stadler, I., Verdier, T., 1991. Strategic managerial incentives and cross ownership structure: a note. *Journal of Economics* 53, 285-297.

Matsumura, T., 1998. Partial privatization in mixed duopoly. *Journal of Public Economics* 70, 473-483.

Matsushima, N., Matsumura, T., 2006. Mixed oligopoly, foreign firms, and location choice. *Regional Science and Urban Economics* 36, 753-772.

Pal, D., 1998. Endogenous timing in a mixed oligopoly. *Economics Letters* 61, 181-185.

Saha, B., Sensarma, R., 2008. The distributive role of managerial incentives in a mixed duopoly. *Economics Bulletin* 12 (27), 1-10.

Simpson, R.D., 1995. Optimal pollution taxation in a Cournot duopoly. *Environmental and Resource Economics* 6, 359-369.

Sklivas, S. D., 1987. The strategic choice of managerial incentives. *Rand Journal of Economics* 18, 452-458.

Tomaru, Y., Saito, M., 2009. Mixed oligopoly, privatization and subsidization in an endogenous timing framework. *The Manchester School* 78, 41-59.

Tomaru, Y., Nakamura, Y., Saito, M., 2009. Subsidization and privatization in a mixed duopoly with managerial delegation. G-COE GLOPE II Working Paper Series, Working Paper No.34, Waseda University.

Ulph, A., 1996. Environmental policy and international trade when governments and producers act strategically. *Journal of Environmental Economics and Management* 30, 265-281.

Vickers, J., 1985. Delegation and the theory of the firm. *Economic Journal Supplement* 95, 138-147.

Wang, L.F.S., Wang, J., 2009. Environmental taxes in a differentiated mixed duopoly. *Economic Systems* 33, 389-396.

Wang, L.F.S., Wang, Y-C., Zhao, L., 2009a. Privatization and the environment in a mixed duopoly with pollution abatement. *Economics Bulletin* 29(4), 3112-3119.

Wang, L.F.S., Wang, Y-C., Zhao, L., 2009b. Market share delegation and strategic trade. *Journal of Industry, Competition and Trade* 9, 49-56.

Wang, L.F.S., Wang, Y-C., Zhao, W., 2008. Strategic trade *Economics Bulletin* 38(6), 1-8.

White, M.D., 2001. Managerial incentives and the decision to hire Managers in markets with public and private firms. *European Journal of Political Economy* 17, 877-896.

In: Economics of Innovation, Incentives and Uncertainty
Editors: P. Simmons and S. T. Jordan
ISBN: 978-1-62257-251-9

Chapter 9

MONETARY UNCERTAINTY AND THE DEMAND FOR MONEY IN THE US

Mohsen Bahmani-Oskooee[1,*] ***and Sahar Bahmani***[2,†]
[1]The Center for Research on International Economics, The Department of Economics, University of Wisconsin-Milwaukee, Milwaukee, WI, US
[2]Department of Economics, University of Wisconsin-Parkside, Kenosha, WI, US

ABSTRACT

The current prolonged recession in the U.S. has necessitated the Fed to engage in expansionary monetary policy, to a degree of which is above and beyond most people's expectation. Such quantitative easing has eroded both consumer and business confidence due to the fear of inflation. The Nobel Laureate Milton Friedman (1984) argued that the increased volatility of money growth raises the degree of perceived uncertainty and therefore, increases the demand for money. This is known as Friedman's "volatility hypothesis".

In this chapter, we revisit the "volatility hypothesis" in the U.S. by estimating the demand for the M2 monetary aggregate. A standard money demand that includes a measure of monetary uncertainty, in addition to income and interest rate, is estimated using recent advances in econometrics literature and annual data over 1971-2010. Friedman's hypothesis is supported, implying that the current prolonged recession in the U.S. could partly be due to the increased holding of cash because of monetary uncertainty.

I. INTRODUCTION

One debate in macroeconomics centers around whether central banks should concentrate on controlling interest rates or target monetary aggregates. Here, in the U.S., the former

* E-mail: bahmani@uwm.edu.
† E-mail: bahmani@uwp.edu.

policy was followed prior to October 1979. In October 1979 the Fed abandoned interest rate policy and shifted to targeting monetary aggregates. During the first three years of the new policy, the volatility of monetary growth in the U.S. tripled. Given the one to one relation between monetary growth and inflation, the fact that the inflation rate declined by more than the money growth rate during 1981-83, relative to 1979-81, is attributed to the volatility of monetary growth by Friedman (1984, p. 399). As he argued, exceptional volatility of monetary growth increases the degree of perceived uncertainty and thereby increases the demand for money.[1]

In testing Friedman's hypothesis, most researchers have followed a traditional approach of including a measure of monetary growth volatility in the specification of the demand for money and tested the significance of this new variable in the presence of other determinants of the demand for money. A significantly positive coefficient will indeed support the Friedman volatility hypothesis. Indeed, the hypothesis is supported for the U.S. by Choi and Oh (2003), who estimated an M1 demand for money which included output, interest rate, financial innovations, output uncertainty and monetary uncertainty. As they show, monetary uncertainty indeed increases the demand for money in the U.S. while output uncertainty reduced it. Bahmani-Oskooee and Xi (2011) who tested the hypothesis for Australia, demonstrated that the outcome could be sensitive to methods of constructing uncertainty measures. When monetary uncertainty was generated using the GARCH method, monetary uncertainty actually had a negative short-run effect, which did not last into the long run. However, when uncertainty was measured by the moving standard deviation method, monetary uncertainty had neither short-run nor long-run effects on the M3 money demand in Australia. Bahmani-Oskooee *et al*. (2012) tested the hypothesis using Chinese data. They also found that monetary uncertainty had only short-run positive effects on the demand for money in China that did not last into the long run. Finally, Bruggemann and Nautz (1997) tested the hypothesis using data from unified Germany. Unlike the above studies, the examination of the volatility-augmented money demand function rejected Friedman's hypothesis by showing that increased money growth volatility decreased, not increased, the demand for money in Germany.

One implication of the volatility hypothesis is that an increased demand for money, due to monetary volatility could induce a decline in consumer spending, which could lead to a decline in output. If this found to be true, this could be one of many reasons for explaining the current prolonged recession in the U.S. Therefore, our purpose in this chapter is to revisit Friedman's hypothesis in the U.S. one more time and determine whether increased volatility in M2 monetary aggregate, during the period covering 1970-2010, has indeed increased the demand for money. To this end, in Section II we specify the demand for M2 monetary aggregate and introduce our methodology. Section III introduces the estimation results with a summary that appears in Section IV. Data definition and sources appear in an Appendix.

[1] Of course, increased demand for money or holding money results in less spending and therefore, a lower inflation rate.

II. The Money Demand Function and Estimation Method

All previous studies that have estimated a money demand function for any country have included in their specification a scale variable, which is measured by the level of economic activity, i.e., real GDP, and a measure of opportunity cost of holding money, which is measured by an interest rate. Thus, we assume that the demand for M2 monetary aggregate in the U.S. takes the following specification:

$$LnM_t = a + bLnY_t + cLnr_t + \varepsilon_t \tag{1}$$

where M is the real M2 monetary aggregate, Y is the real GDP, and r is an interest rate. The transaction demand for money postulates that an increase in the level of economic activity, or Y, induces an increase in the demand for money due to the volume of goods and services produced. Therefore, we expect an estimate of b to be positive. On the other hand, since the interest rate measures the opportunity cost of holding money, an increase in it lowers the demand for money, or cash holding, implying that an estimate of c should be negative.

Since our goal in this chapter is to assess the volatility of a monetary aggregate on the demand for money, we extend specification (1) by adding a volatility measure of money as another determinant of the money demand in the U.S.

$$LnM_t = a + bLnY_t + cLnr_t + dLnVM_t + \varepsilon_t \tag{2}$$

where VM is a measure of volatility of a monetary aggregate. Friedman's volatility hypothesis implies that an estimate of d should be positive.

The estimation of equation (2) by any mean at all, only yields long-run coefficient estimates. However, it is possible for some of the variables to have short-run effects but not long-run effects. Given the advances in econometrics literature, one way to asses the short-run effects of the right-hand side variables in equation (2) is to include the short-run adjustment mechanism by specifying (2) in an error-correction modeling format. To this end, we follow Pesaran *et al.'s* (2001) bounds testing approach and rewrite equation (2) in an error-correction modeling format as in (3):[2]

$$\Delta LnM_t = \alpha + \sum_{i=1}^{n1} \beta_i \Delta LnM_{t-i} + \sum_{i=0}^{n2} \delta_i \Delta LnY_{t-i} + \sum_{i=0}^{n3} \phi_i Lnr_{t-i} + \sum_{i=0}^{n4} \mu_i \Delta LnVM_{t-i}$$

$$+ \rho_0 LnM_{t-1} + \rho_1 LnY_{t-1} + \rho_2 Lnr_{t-1} + \rho_3 LnVM_{t-1} + \varepsilon_t \tag{3}$$

Note that in specification (3), the linear combination of lagged level variables are included as a proxy for the lagged error term from (2). How do we justify the inclusion of these lagged level variables in a standard VAR model? Pesaran *et al.* (2001) propose applying

[2] The methodology here closely follows Bahmani-Oskooee and Tanku (2008) and Bahmani-Oskooee and Gelan (2009).

the familiar F test to establish their joint significance which will also serve to be a method of establishing cointegration among the variables. However, unlike the standard F test, in this application they tabulate new critical values because of different integrating properties of the variables involved. By assuming all variables in a given model to be integrated of order one, they tabulate, through a Monte Carlo experiment, an upper bound critical value at different significance levels, which also depend on the number of regressors. A lower bound critical value is provided when all variables are stationary or integrated of order zero.[3] Pesaran *et al.* (2001) show that the upper bound critical value could still be used to establish cointegration among the variables, even if some variables are integrated of order one and some are integrated of order zero. Since macro time-series variables are rarely integrated of an order beyond one, under this method there is no need for pre unit-root testing of variables.

Once cointegration among the variables is established, the long-run effects are inferred by the estimates of ρ_1 - ρ_3 which are normalized on ρ_0. This normalization basically requires setting the following portion of the estimates equal to zero, as in (4):

$$\hat{\rho}_0 LnM_{t-1} + \hat{\rho}_1 LnY_{t-1} + \hat{\rho}_2 Lnr_{t-1} + \hat{\rho}_3 LnVM_{t-1} = 0 \tag{4}$$

Solving (4) for $Ln\ M_{t-1}$ yields:

$$LnM_{t-1} = -\frac{\hat{\rho}_1}{\hat{\rho}_0} LnY_{t-1} - \frac{\hat{\rho}_2}{\hat{\rho}_0} Lnr_{t-1} - \frac{\hat{\rho}_3}{\hat{\rho}_0} LnVM_{t-1} \tag{5}$$

Short-run effects of each variable are inferred by the estimate of coefficients attached to first-differenced coefficients of that variable. For example, short-run effects of volatility of money are inferred by the estimates of μ_i's. Note that in addition to avoiding pre unit-root testing, another advantage of this method is that the short-run as well as the long-run effects are estimated in one step by applying the Ordinary Least Square method to equation (3), which is the subject of estimation in the next section.

III. The Results

In this section, we estimate the error-correction model (3) using the U.S. annual data over the period 1970-2010. A common practice in estimating any VAR model is to use a set criterion in selecting the optimum number of lags on each first-differenced variable. Here, we impose a maximum of four lags on each variable and use Akaike's Information Criterion (AIC) to select the optimum lags.

We then report the results of the optimum model. Since the model includes short-run as well as long-run components, the important question is what kind of interest rate should be used. To this end, we use the government 10-year bond yield first and the T-Bill rate next. Using the bond yield, we report the results in Table 1.

[3] A variable is said to be integrated of order d if it achieves stationarity after being differenced d times.

Table 1. Full-Information Estimate of Equation (3)

Panel A: Short-Run Coefficient Estimates						
	Lag Order					
	0	1	2	3	4	5
Δ Ln M	-	0.4185				
		(4.15)				
Δ Ln Y	0.0663	-0.0923				
	(2.39)	(3.74)				
ΔLn r	-0.0208					
	(4.51)					
Δ Ln VM	0.0095					
	(6.95)					
Panel B: Long-Run Coefficient Estimates						
Constant	*Ln Y*	*Ln r*	*Ln VM*			
-0.0015	0.1272	-0.0061	0.0594			
(0.01)	(3.92)	(0.33)	(3.10)			
Panel C: Diagnostic Statistics						
F	ECM_{t-1}	*LM*	RESET	*Adj.* R^2		
7.75	-0.1331	0.29	5.78	0.83		
	(5.64)					

Notes: a. Numbers inside parentheses are absolute value of t-ratios.

b. The upper bound critical value of the F statistic at the usual 5% level of significance is 4.01. This comes from Pesaran *et al.* (2001, Table CI-Case III, p. 300).

c. LM is the Lagrange multiplier test for serial correlation. It has a χ^2 distribution with 1 degree of freedom. The critical value at the 5% level of significance is 3.84.

d. RESET is Ramsey's test for functional form. It is also distributed as $\chi^2_{(1)}$.

There are three panels in Table 1. While Panel A reports the short-run coefficient estimates, Panel B reports the long-run estimates. Diagnostic statistics are reported in Panel C. From Panel A, it is evident that all short-run coefficient estimates are highly significant. Concentrating on our variable of interest, we gather that the volatility measure of money (*Ln MV*) carries its expected positive sign and highly significant coefficient, implying that monetary uncertainty results in an increase in the demand for money in the U.S., confirming again, Friedman's volatility hypothesis. Do these short-run effects translate into the long run? The results appear in Panel B.

From Panel B it is clear that short-run effects of income and monetary volatility indeed last into the long run. Both variables do carry highly significant coefficients with their expected signs. Thus, it appears that monetary uncertainty has not only short-run effects on the demand for money in the U.S., but also it has long-run effects. However, these long-run effects will not be spurious if we only establish joint significance of lagged level variables, or cointegration among them. To this end, we shift to Panel C.

The results of the F test along with a few other diagnostics do appear in Panel C. Clearly, our calculated F statistic is much higher than its critical value of 3.61, supporting cointegration among the four variables. The Lagrange Multiplier (LM) statistic is also reported. This statistic is used to test autocorrelation among the residuals of the optimum model. It has a χ^2 distribution with one degree of freedom. Since our calculated statistic is much less than its critical value of 3.84, there is no evidence of serial correlation among the residuals. Ramsey's RESET test for functional misspecification is also reported. This statistic also has a χ^2 distribution with one degree of freedom. Our calculated statistic is somewhat higher than its critical value, hinting to a slightly misspecified model, which we will resolve later.

Two additional questions about the estimates of any money demand function are in order. The first is whether the adjustment of the variables is toward their long-run equilibrium values. Within the bounds testing approach this is answered by using the long-run normalized coefficients and equation (5) to calculate a lagged error-correction term, ECM_{t-1}. Then the linear combination of the lagged level variables in (3) is replaced by ECM_{t-1}, as in equation (6):

$$\Delta LnM_t = \alpha + \sum_{i=1}^{n1} \beta_i \Delta LnM_{t-i} + \sum_{i=0}^{n2} \delta_i \Delta LnY_{t-i} + \sum_{i=0}^{n3} \phi_i Lnr_{t-i} + \sum_{i=0}^{n4} \mu_i \Delta LnVM_{t-i}$$

$$+ \rho ECM_{t-1} + \varepsilon_t \tag{6}$$

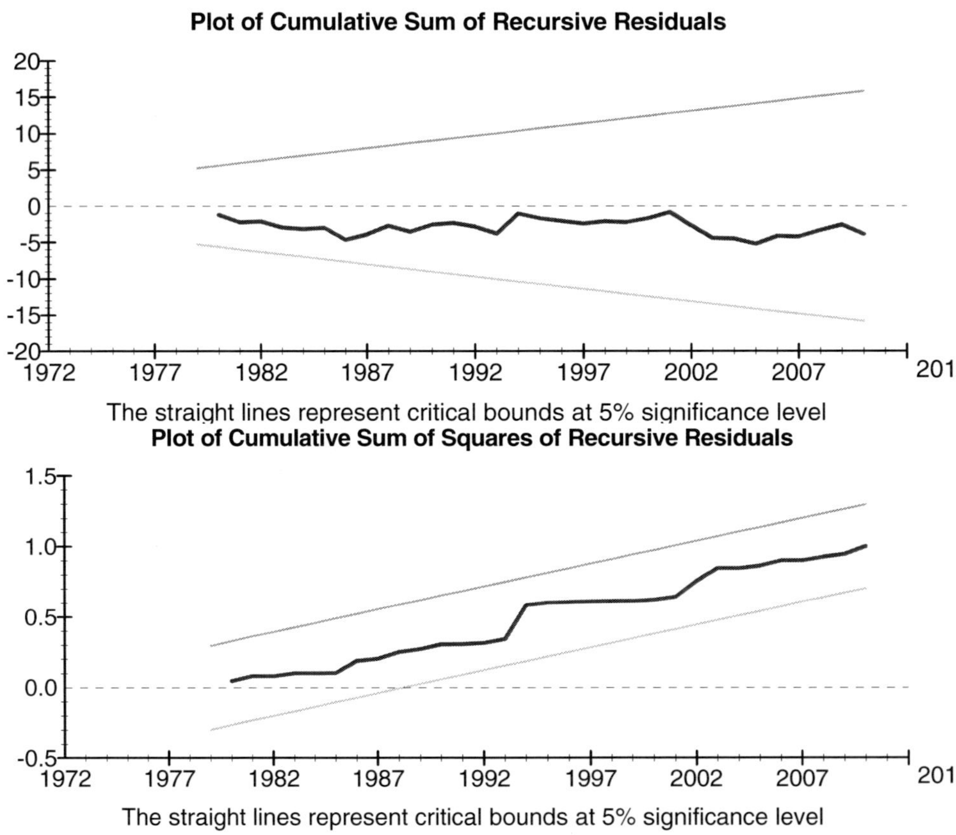

Figure 1. Plot of CUSUM and CUSUMSQ Tests.

We then estimate equation (6) one more time after imposing the same optimum lags on each first-differenced variable, i.e., after setting n1=n2=n3=n4=1. If adjustment is to be toward long-run equilibrium, ECM_{t-1} must carry a significantly negative coefficient. Indeed, the results reported in Panel C confirm adjustment toward long-run equilibrium since the estimated coefficient of ECM_{t-1} is negative and highly significant. At this stage, we also apply the well-known CUSUM and CUSUMSQ tests to the residuals of equation (6). These tests are proposed by Brown, Durbin and Evans (1975) to test for stability of the short run as well as the long run coefficient estimates. Applying these tests usually require that we break the sample in such a way that there are enough observations before and after the break point. By moving one observation from one period to the other, the two statistics are re-calculated and then plotted against the break points, or break dates. If all the coefficients are to be stable, the plot of these statistics should stay within the 5% significance level, which is portrayed by two straight lines, whose equations are given in Brown *et. al* 1975, section 2.3. A common practice is to adhere to a graphical presentation of these tests, as in Figure 1. Clearly, both tests support the stability of the estimated money demand function for the U.S.

As mentioned before, we do some sensitivity analysis by replacing the government bond yield with the Treasury bill rate. We then follow the same estimation procedure and report the results in Table 2.

Table 2. Full-Information Estimate of Equation (3) (when T-Bill rate is used)

Panel A: Short-Run Coefficient Estimates						
			Lag Order			
	0	1	2	3	4	5
Δ Ln M	-	-				
Δ Ln Y	0.1755 (5.73)					
ΔLn r	-0.0085 (6.10)					
Δ Ln VM	0.0105 (8.78)					
Panel B: Long-Run Coefficient Estimates						
Constant	*Ln Y*	*Ln r*	*Ln VM*			
0.0618 (0.22)	0.1008 (2.19)	-0.0002 (0.02)	0.0944 (2.39)			
Panel C: Diagnostic Statistics						
F	ECM_{t-1}	*LM*	RESET	*Adj.* R^2		
10.74	-0.1120 (2.45)	2.08	3.43	0.76		

Notes: a. Numbers inside parentheses are absolute value of t-ratios.

b. The upper bound critical value of the F statistic at the usual 5% level of significance is 4.01. This comes from Pesaran *et al.* (2001, Table CI-Case III, p. 300).

c. LM is the Lagrange multiplier test for serial correlation. It has a χ^2 distribution with 1 degree of freedom. The critical value at the 5% level of significance is 3.84.

d. RESET is Ramsey's test for functional form. It is also distributed as $\chi^2_{(1)}$.

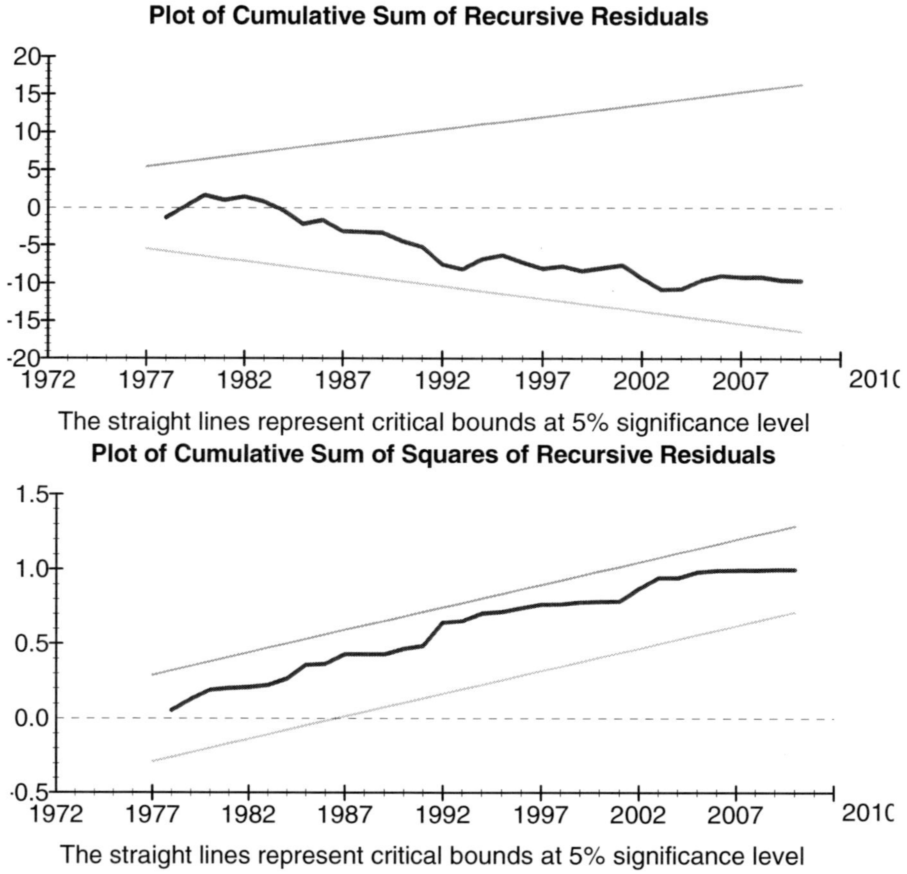

Figure 2. Plot of CUSUM and CUSUMSQ Tests (when T-bill rate is used in the model).

From Panel A of Table 2, clearly all three variables do have short-run significant effects on the demand for money. However, only income and money volatility have long-run effects. Again, the long-run results support Friedman's volatility hypothesis. Once again, from Panel C, further support for cointegration is provided by the highly significant F test. The LM statistic, again, supports correlation free residuals and this time, Ramsey's RESET test is insignificant, supporting a correctly specified optimum model. Furthermore, the results of the CUSUM and CUSUMSQ tests, as reflected in Figure 2, support a stable demand for M2 monetary aggregate in the U.S.[4]

Conclusion

When the Fed switched from fixing interest rates to targeting monetary aggregates in October 1979, the inflation rate in the U.S. declined at a rate more than the decline in the growth rate of the money supply, which was in violation of the monetarist rule. Friedman defended monetarists by arguing that during the post 1979 period, the volatility of money increased and this made the public to be cautious, leading to an increase in their demand for

[4] For the additional application of these tests, see Bahmani (2008).

money. Since people held more money, they had less for spending which resulted in a lower inflation rate. The Friedman hypothesis has been tested either by investigating the impact of increased volatility on velocity of money or directly on the demand for money.

In this chapter, we chose the later approach and included a measure of volatility of M2 monetary aggregate in the demand for money function. The model is tested for the U.S. using annual data over the period 1970-2010. We use a method that does not require pre unit root testing and variables could be stationary or non-stationary, i.e., the bounds testing approach to cointegration and error-correction modeling. Furthermore, this method allows us to distinguish short-run effects from long-run effects. No matter what sort of interest rate we used in formulating our demand for money, the volatility measure of the M2 monetary aggregate is shown to have short-run as well as long-run effects.

APPENDIX
DATA DEFINITION AND SOURCES

All data are annual over the period 1970-2010 and collected from the International Financial Statistics of IMF.

Variables:

M2 = Money aggregate defined as real M2. Nominal M2 figures are deflated by GDP deflator.

Y: Real GDP.

r = interest rate defined as either the T-Bill rate or Government bond yield.

VM = Volatility measure of nominal M2. For each year it is defined as the standard deviation of 12 monthly nominal M2 values within that year. This approach is followed by others in other areas as well. For example, Bahmani-Oskooee and Hegerty (2009) who constructed the volatility measure of the real exchange rate, relied upon this definition.

REFERENCES

Bahmani, S., (2008) "The Stability of the Demand for Money in the Middle East," *Emerging Markets Finance and Trade*, Volume 44, pp. 62-83.

Bahmani-Oskooee, M. and A. Tanku, (2008), "The Black Market Exchange Rate vs. the Official Rate in Testing PPP: Which Rate Fosters the Adjustment Process", *Economics Letters,* Vol. 99, pp. 40-43.

Bahmani-Oskooee, M., and S. W. Hegerty (2009), "The Effects of Exchange-Rate Volatility on Commodity Trade Between the U.S. and Mexico," *Southern Economic Journal,* Vol. 79, pp. 1019-1044.

Bahmani-Oskooee, M. and A. Gelan, (2009), . "How Stable is the Demand for Money in African Countries", *Journal of Economic Studies*, Vol. 36, pp. 216-235.

Bahmani-Oskooee, M. and D. Xi, (2011), "Economic Uncertainty, Monetary Uncertainty and the Demand for Money in Australia" *Australian Economic Papers,* Vol. 50, pp. 115-128.

Bahmani-Oskooee, M. and Y. Wang (2012), “Economic Uncertainty, Monetary Uncertainty and the Demand for Money in China”, *The Chinese Economy*, forthcoming.

Brown, R. L., J. Durbin, and J. M. Evans (1975), "Techniques for Testing the Constancy of Regression Relations Over Time", *Journal of the Royal Statistical Society*, Series B, Vol. 37, pp. 149-163.

Bruggemann, I. and D. Nautz (1997), “Money Growth Volatility and the Demand for Money in Germany: Friedman’s Volatility Hypothesis Revisited”, Weltwirtschaftliches Archiv, Vol. 133, pp. 523-537.

Choi, W. G., and S. Oh, (2003), “A Demand Function with Output Uncertainty, Monetary Uncertainty, and Financial Innovations”, *Journal of Money, Credit, and Banking,* Vol. 35, pp. 685-709.

Friedman, M. (1984), “Lessons from the 1979-1982 Monetary Policy Experiment”, *American Economic Review, Papers and Proceedings*, Vol. 74, pp. 397-400.

Pesaran, M. H.’ Y. Shin, and R. J. Smith, (2001), “Bounds Testing Approaches to the Analysis of Level Relationships,” *Journal of Applied Econometrics,* Vol.16, pp. 289-326.

In: Economics of Innovation, Incentives and Uncertainty ISBN: 978-1-62257-251-9
Editors: P. Simmons and S. T. Jordan

Chapter 10

AN EFFICIENT BONUS CONTRACT

***Susheng Wang*[*]**
Department of Economics, Hong Kong University of Science and Technology, Hong Kong

ABSTRACT

Why are bonus/promotion schemes so widely used in reality? Are they effective in alleviating incentive problems? For the standard agency model, this paper proposes an alternative solution to the classical solution in Holmström (1979). The advantages of our solution are that (a) it is a simple solution; (b) it is not based on the troublesome first-order approach; and (c) it is the first best. The disadvantage of our solution is that it imposes a boundary condition on the support of the distribution function. However, this boundary condition does have sensible economic interpretations and our optimal contract resembles the widely observed bonus/promotion scheme in reality. Hence, our solution shows the potential of a bonus/promotion scheme in resolving incentive problems.

Keywords: One-step contract, Bonus mechanism

JEL Classification: D23

INTRODUCTION

The objective of this paper is to explain why bonus/promotion schemes are so popular in reality. We do this by finding an alternative optimal contract for the standard agency model. As we known, when there is no risk aversion, there is a simple optimal contract: a *linear*

[*] E-mail: s.wang@ust.hk.

contract.[1] Linear contracts are popular in reality. However, when there is risk aversion, the optimal contract becomes very complex, which severely prevents its use in applications. This paper finds two simple conditions under which the standard agency model with risk aversion has a simple optimal contract: a *bonus contract*. This contact is convenient for applications and it is implementable by a simple bonus/promotion scheme.

Real-world contracts between principals and agents are typically very simple and often have a multi-step bonus structure that specifies wage increases for certain minimum levels of performance. Why can contracts be so simple in reality? This paper attempts to provide an answer by establishing the optimality of a one-step contract for the standard agency model. Our optimal contract is equivalent to a bonus scheme by which an agent is rewarded a bonus upon achieving a minimum performance. Such a bonus, reward or promotion scheme is widely observed in reality and is generally believed to be a key incentive device.

The classical solution to the standard principal-agent model is based on the *first-order approach (FOA)*. The foundation of this solution is laid down by Mirrlees (1999) and Holmström (1979). However, due to its complex structure, the existence of the solution (Mirrlees 1999 and Holmström 1979), the validity of the FOA (Rogerson 1985 and Jewitt 1988) and the monotonicity of the solution (Grossman–Hart 1983) are all difficult issues. As a result, the classical solution remains elusive to those applied fields where risk aversion is a necessity. In contrast, our bonus contract does not face these difficulties. In fact, we find simple conditions under which (a) the solution is globally optimal and (b) the *incentive-compatibility (IC) condition* is satisfied either locally or globally.

This paper proceeds as follows. We present the agency model in the next section. The solution is derived in Section 3. Discussion about our solution is made in Section 4. Two examples are presented in Section 5. Finally, we conclude the paper with some remarks in Section 6.

THE MODEL

Let a be the agent's effort and $\mathbb{A}$ be the effort space. Let the contract space be

$$\mathcal{S} \equiv \{ s : \mathbb{R} \to \mathbb{R}_+ \mid s \text{ is Lesbegue integrable} \},$$

where we have implicitly imposed the limited liability condition: $s(x) \geq 0$, just as Mirrlees (1999) and Holmström (1979) did. Let the density function of output be $f(x,a)$ for $A(a) \leq x \leq B(a)$ and zero elsewhere, where the domain is $\mathcal{D}(f) \equiv [A(a),\ B(a)]$. We allow $B(a)$ to be arbitrary, including being a constant or $B(a) = +\infty$. But, we will impose a condition on $A(a)$ later. Given a utility function $u : \mathbb{R}_+ \to \mathbb{R}_+$, the agent's expected utility is

[1] Bhattacharyya–Lafontaine (1995) are the first to provide such a result for an additive output process of the form $x = a + \varepsilon$. Kim–Wang (1998) provide a general result.

$$U(a,s) \equiv \int_{A(a)}^{B(a)} u[s(x)]f(x,a)dx - c(a).$$

Given a contract $s(\cdot)$, the agent's problem is

$$\max_{a \in \mathbb{A}} \int_{A(a)}^{B(a)} u[s(x)]f(x,a)dx - c(a).$$

Thus, the principal's problem is

$$\begin{aligned}
\pi \equiv \max_{s \in \mathcal{S}, a \in \mathbb{A}} & \int_{A(a)}^{B(a)} [x - s(x)]f(x,a)dx \\
\text{s.t.} \quad IR: & \int_{A(a)}^{B(a)} u[s(x)]f(x,a)dx \geq c(a), \qquad (1) \\
IC: & \ a \in \arg\max_{a'} U(a', s).
\end{aligned}$$

Here, the reservation value is taken to be $\overline{u} = 0$. If $\overline{u} \neq 0$, we can replace $c(a)$ by $C(a) \equiv c(a) + \overline{u}$ and the model remains the same.

This model is the same as the standard agency model formulated in Holmström (1979) except that we allow the boundaries of the domain to be dependent on effort. It turns out that this dependence is important for our alternative solution to the classical solution in Holmström (1979). Another difference is that we do not rely on the FOA.

Assumption 1. The utility function $u(x)$ is concave, onto and strictly increasing.

Assumption 2. The expected revenue $R(a) \equiv \int_{A(a)}^{B(a)} xf(x,a)dx$ is increasing and concave.

Assumption 3. (FOSD). $F_a(x,a) \leq 0$ for any $x \in \mathbb{R}$ and $a \in \mathbb{A}$.

These three assumptions are natural requirements. In particular, u being onto and strictly increasing is to ensure $u^{-1}[c(a)]$ to be well defined, and Assumption 3 is necessary for any sensible contract theory.

THE SOLUTION

We now derive the solution of model (1). Given optimal contract $s^*(x)$ together with optimal effort a^*, as shown in Section 3.3 later, $U(a, s^*)$ is generally not continuously differentiable at $a = a^*$. Hence, the first-order condition $U_a(a^*, s^*) = 0$ cannot replace the IC condition in problem (1). This means that the FOA is no longer applicable to our model. We need to use a new approach for this model. Our approach is to consider the first-best version of problem (1) by ignoring the IC condition first and later to show that the first-best solution is also the solution of problem (1).

Proposition 1. *Let* a^* *be the solution of the following equation:*

$$\frac{\partial u^{-1}[c(a^*)]}{\partial a} = R'(a^*), \tag{2}$$

and suppose that the following two conditions are satisfied:

$$\frac{\partial \ln u^{-1}[c(a^*)]}{\partial a} \geq \frac{\partial \ln c(a^*)}{\partial a}, \tag{3}$$

$$-F_a[A(a^*), a^*] > \frac{\partial \ln c(a^*)}{\partial a}. \tag{4}$$

Then, under Assumptions 1–3, the optimal effort is a^* *and the optimal contract is*

$$s^*(x) = \begin{cases} 0 & \text{if} \quad x < A(a^*), \\ u^{-1}[c(a^*)] & \text{if} \quad x \geq A(a^*). \end{cases} \tag{5}$$

Furthermore, (a^*, s^*) *achieves the first best.*

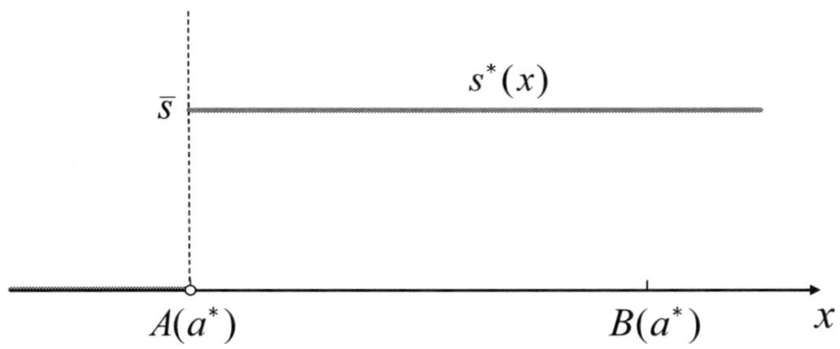

Proof. We proceed the proof by three steps.

Step 1. The Optimal Contract
Without the IC condition, the Lagrangian for problem (1) is

$$L \equiv \int_{A(a)}^{B(a)} [x - s(x)] f(x,a) dx + \lambda \left[\int_{A(a)}^{B(a)} u[s(x)] f(x,a) dx - c(a) \right],$$

where $\lambda \geq 0$ is a Lagrange multiplier. The Hamiltonian is

$$H(x,s) \equiv x - s + \lambda u(s).$$

The Euler equation $H_s = 0$ implies

$$. \lambda u'[s(x)] = 1. \tag{6}$$

Thus, by Pontryagin's maximum principle, the optimal contract is

$$s^*(x) = \begin{cases} 0 & \text{if } x < A(a^*), \\ \overline{s} & \text{if } x \geq A(a^*), \end{cases}$$

where $u'(\overline{s}) \equiv 1/\lambda$. Note that a jump at $A(a^*)$ is necessary for satisfying the IC condition, as shown later in Section 3.3. By the IR condition in (1), we have

$$c(a^*) = \int_{A(a^*)}^{B(a)} u(\overline{s}) f(x,a^*) dx = u(\overline{s}),$$

implying

$$. \overline{s} = u^{-1}[c(a^*)]. \tag{7}$$

Here, $u(x)$ being onto and strictly increasing ensures the existence and uniqueness of $\overline{s}$. Thus,

$$. s^*(x) = \begin{cases} 0 & \text{if } x < A(a^*), \\ u^{-1}[c(a^*)] & \text{if } x \geq A(a^*). \end{cases} \tag{8}$$

Note that this optimal contract is dependent on the optimal effort a^*, which is determined in the next step.

Step 2. The Optimal Effort

Given $s^*(x)$ in (8), the Lagrangian becomes

$$. L(a) = \int_{A(a)}^{B(a)} x f(x,a) dx - \int_{A(a)}^{B(a)} s^*(x) f(x,a) dx + \lambda \left[\int_{A(a)}^{B(a)} u[s^*(x)] f(x,a) dx - c(a) \right].$$

If $a < a^*$,

$$L(a) = R(a) - \bar{s}\int_{A(a^*)}^{B(a)} f(x,a)dx + \lambda\left[u(\bar{s})\int_{A(a^*)}^{B(a)} f(x,a)dx - c(a)\right]$$

$$= R(a) - \bar{s}[1 - F(A(a^*),a)] + \lambda u(\bar{s})[1 - F(A(a^*),a)] - \lambda c(a).$$

Thus,

$$L'(a) = R'(a) - \lambda c'(a) + F_a[A(a^*),a]\left\{u^{-1}[c(a^*)] - \lambda c(a^*)\right\}, \quad \text{for } a < a^*. \tag{9}$$

If $a \geq a^*$,

$$L(a) = R(a) - \bar{s}\int_{A(a)}^{B(a)} f(x,a)dx + \lambda\left[u(\bar{s})\int_{A(a)}^{B(a)} f(x,a)dx - c(a)\right]$$

$$= R(a) - \bar{s} + \lambda\left[u(\bar{s}) - c(a)\right].$$

Thus,

$$L'(a) = R'(a) - \lambda c'(a), \quad \text{for } a > a^*. \tag{10}$$

To ensure the optimality of a^*, let

$$R'(a^*) - \lambda c'(a^*) = 0. \tag{11}$$

Since $R(a) - \lambda c(a)$ is concave in a or $R'(a) - \lambda c'(a)$ is decreasing in a, (11) implies $L'(a) \leq 0$ for $a \geq a^*$. Suppose

$$u^{-1}[c(a^*)] - \lambda c(a^*) \leq 0. \tag{12}$$

Then, by Assumption 3 and the decreasingness of $R'(a) - \lambda c'(a)$, (11) implies $L'(a) \geq 0$ for $a \leq a^*$. In other words, conditions (11) and (12) imply the global optimality of a^*. Notice that $L(a)$ is indeed continuous at a^*. By (6) and (7),

$$\frac{1}{\lambda} = u'\left\{u^{-1}[c(a^*)]\right\}.$$

By (11), this becomes

$$\frac{c'(a^*)}{R'(a^*)} = u'\left\{u^{-1}[c(a^*)]\right\}. \tag{13}$$

Since

$$\frac{\partial u^{-1}[c(a)]}{\partial a} = \frac{c'(a)}{u'\{u^{-1}[c(a)]\}},$$

(13) becomes

$$\frac{\partial u^{-1}[c(a^*)]}{\partial a} = R'(a^*). \tag{14}$$

This equation determines a^*. Also, condition (12) can be written as

$$\frac{c(a^*)}{u^{-1}[c(a^*)]} \geq u'\{u^{-1}[c(a^*)]\},$$

implying

$$\frac{\partial \ln u^{-1}[c(a^*)]}{\partial a} \geq \frac{\partial \ln c(a^*)}{\partial a}. \tag{15}$$

In summary, we need equations (14) and (15), where equation (14) determines a^* and condition (15) ensures the global optimality of a^*.

Step 3. The IC Condition

We now verify the IC condition in (1). For the optimal contract in (8), if $a < a^*$,

$$\begin{aligned} U(a,s^*) &= \int_{A(a^*)}^{B(a)} u(\bar{s}) f(x,a)dx - c(a) = c(a^*)\int_{A(a^*)}^{B(a)} f(x,a)dx - c(a) \\ &= c(a^*)\big[1 - F(A(a^*),a)\big] - c(a), \end{aligned}$$

implying

$$U_a(a,s^*) = -c(a^*)F_a[A(a^*),a] - c'(a), \quad \text{for } a < a^*,$$

implying

$$U_a(a_-^*,s^*) = -c(a^*)F_a[A(a^*),a^*] - c'(a^*).$$

If $a \geq a^*$,

$$U(a,s^*) = \int_{A(a)}^{B(a)} u(\bar{s}) f(x,a)dx - c(a) = c(a^*)\int_{A(a)}^{B(a)} f(x,a)dx - c(a) = c(a^*) - c(a).$$

We have $U(a^*,s^*)=0$ and $U(a,s^*)<U(a^*,s^*)$ when $a>a^*$. Thus, for a^* to be locally optimal, we need to find a $\delta>0$ such that $U(a,s^*)<0$ when $a\in(a^*-\delta,\ a^*)$. This is guaranteed by $U_a(a^*_-,s^*)>0$ or

$$-F_a[A(a^*),a^*]>\frac{c'(a^*)}{c(a^*)}. \tag{16}$$

Notice that $U(a,s^*)$ is indeed continuous at a^*. In other words, under condition (16), given contract $s^*(x)$, a^* is indeed a local maximum point of $U(a,s^*)$. That is, the IC condition is satisfied.

The required conditions (14), (15) and (16) are listed in the proposition. The proof is now complete. *Q.E.D.*

DISCUSSION

We have now shown that a bonus contract is optimal for the standard agency model with risk aversion. As the bonus is $u^{-1}[c(a)]$, condition (2) means that the marginal bonus is equal to the marginal revenue. Condition (3) means that the marginal elasticity of bonus is larger than the marginal elasticity of cost. And, condition (4) means that the marginal chance of getting the bonus is higher than the marginal elasticity of cost.

Although the contract is derived as a revenue-sharing contract, it is implementable by a simple a bonus or promotion scheme. A labor contract in reality often has such a component. In reality, a contract may contain two reward schedules $s_b(x)$ and $s_m(x)$, where $s_b(x)$ is a bonus scheme that may be firm-specific and $s_m(x)$ is the market wage that depends on marketable skills. We have only derived the firm-specific bonus scheme in this paper. The market wage is a separate issue, which will relate to market conditions.

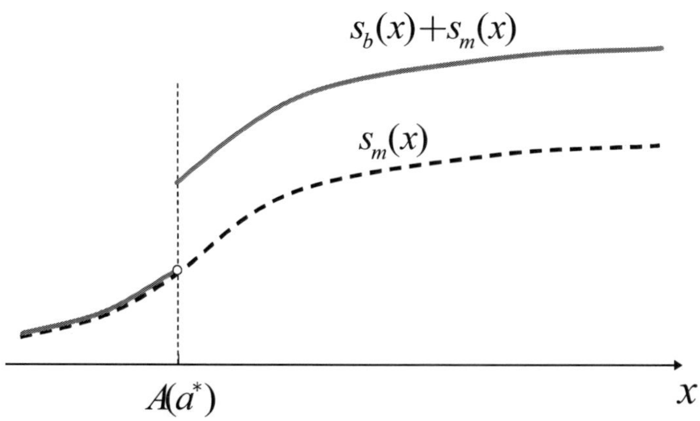

Our solution looks strikingly different from the FOA-based classical solution in Holmström (1979). Our solution looks puzzling at first glance because of its simplicity. However, the intuition for the solution is actually fairly simple. To induce the optimal effort a^*, the FOA suggests that the principal should associate each output level to a payment. Alternatively, our bonus approach suggests that the principal can induce a^* by simply offering a bonus at a minimum level of performance. For the latter strategy to work, the dependence of the distribution function on effort at the minimum level of performance is crucial. Condition (4) is precisely for such dependence.

There are some interesting features in our solution. First, it is a closed-form solution with a clear bonus component. In fact, the bonus is paid at a minimum level of performance. This bonus can be in the form of a promotion, a bonus, a reward, or a change of nature of the contract. Second, the solution has an interesting form with simple and intuitive expressions determining the pay and effort. Third, the solution achieves the first best.

We have some further technical remarks about the conditions in the proposition. First, since both $c(\cdot)$ and $u^{-1}(\cdot)$ are convex, $u^{-1}[c(a)]$ is convex in a, implying that $\frac{\partial u^{-1}[c(a)]}{\partial a}$ is increasing in a. Also, since $R'(a)$ is decreasing in a, the optimal effort a^* from (2) must be unique.

Second, since u^{-1} is convex, $u^{-1}[c(a)]$ is more convex than $c(a)$ is. This indicates that condition (3) can be easily satisfied for any $a \in \mathbb{A}$. For example, for $u(x) = x^{1-\alpha}$ and $c(a) = a^{\beta}$, where $\alpha \in [0, 1]$ is the relative risk aversion and $\beta \geq 1$, (3) is satisfied for any $a \in \mathbb{R}_+$.

Third, there is no requirement on the right boundary $B(a)$ at all.

Fourth, by the same derivation as in the proof, a^* is the global maximum point of $U(a, s^*)$ if (16) is replaced by the following stronger version:

$$-F_a[A(a^*), a] \geq \frac{c'(a)}{c(a^*)}, \quad \text{for any } a < a^*. \tag{17}$$

Finally, we have

$$F_a[A(a), a] = -f[A(a), a]A'(a), \quad \text{for any } a \in \mathbb{A}. \tag{18}$$

Equation (2) indicates that a^* is independent of many features in the distribution function. For example, if $x = A(a) + \varepsilon$, where $\varepsilon \geq 0$ is a random variable with mean $\bar{\varepsilon}$, and if $F_{\varepsilon}(x)$ and $f_{\varepsilon}(x)$ are respectively the distribution and density functions of ε, then $F(x, a) = F_{\varepsilon}[x - A(a)]$, implying

$$R(a) = A(a) + \bar{\varepsilon}, \quad -F_a[A(a), a] = f_\varepsilon(0)A'(a).$$

Then, equation (2) and condition (4) respectively become

$$\frac{\partial u^{-1}[c(a^*)]}{\partial a} = A'(a^*), \tag{19}$$

$$f_\varepsilon(0) > \frac{1}{A'(a^*)} \frac{\partial \ln c(a^*)}{\partial a}. \tag{20}$$

From (19), we can see that a^* has nothing to do with the distribution function $F_\varepsilon(x)$. This means that the right-hand side of (20) is independent of the distribution function. Hence, condition (20) defines the class of density functions $f_\varepsilon(x)$ on $[0, \infty)$ that has a lower bound at the left boundary. For example, if ε follows the exponential distribution with $f_\varepsilon(x) = \frac{1}{\sigma} e^{-\frac{x}{\sigma}}$ for $x \geq 0$, where $\sigma > 0$ is the standard variance, then we have $f_\varepsilon(0) = \frac{1}{\sigma}$, which satisfies (20) when σ is sufficiently small.

Examples

We consider two examples, which use the exponential distribution and the uniform distribution, respectively.

Example 1.

Suppose

$$u(x) = x^{1-\alpha},$$

$$c(a) = a^\beta,$$

$$f(x, a) = \frac{1}{a} e^{-\frac{x-A(a)}{a}}, \quad \text{for } x \geq A(a),$$

where $A(a)$ is an arbitrary increasing function, $\alpha \in [0, 1]$ is the relative risk aversion, and $\beta \geq 1$. This set of functions includes the example in Holmström (1979) as a special case. We have

$$u^{-1}[c(a)] = a^{\frac{\beta}{1-\alpha}}, \quad R(a) = A(a) + a, \quad F_a[A(a), a] = -\frac{A'(a)}{a}.$$

Condition (2) becomes

$$\frac{\beta}{1-\alpha}(a^*)^{\frac{\beta}{1-\alpha}-1}=1+A'(a^*). \tag{21}$$

Condition (3) is satisfied for any $a \geq 0$. Condition (4) becomes

$$\beta < A'(a^*). \tag{22}$$

Let $A(a)=\gamma a$ for some $\gamma > 1$. Then, (22) is satisfied for any $a \in \mathbb{A}$ and for $\beta \leq \gamma$; in fact, the stronger version of (4), condition (17), is also satisfied. By (21), we find

$$a^* = \left[\frac{(1+\gamma)(1-\alpha)}{\beta}\right]^{\frac{1-\alpha}{\alpha+\beta-1}}, \qquad \overline{s} = \left[\frac{(1+\gamma)(1-\alpha)}{\beta}\right]^{\frac{\beta}{\alpha+\beta-1}}.$$

Example 2.
Suppose

$$u(x)=x^{1-\alpha},$$
$$c(a)=a^{\beta},$$
$$f(x,a)=\frac{1}{\sigma}, \quad \text{for } A(a)\leq x \leq A(a)+\sigma,$$

where $A(a)$ is an arbitrary increasing function, $\alpha \in [0,\ 1]$ is the relative risk aversion, $\sigma > 0$, and $\beta \geq 1$. We have

$$u^{-1}[c(a)]=a^{\frac{\beta}{1-\alpha}}, \quad R(a)=A(a)+\frac{\sigma}{2}, \quad F_a[A(a),a]=-\frac{1}{\sigma}A'(a).$$

Condition (2) becomes

$$\frac{\beta}{1-\alpha}(a^*)^{\frac{\beta}{1-\alpha}-1}=A'(a^*). \tag{23}$$

Condition (3) is satisfied for any $a \geq 0$. Condition (4) becomes

$$\sigma < \frac{1}{\beta}a^* A'(a^*), \tag{24}$$

which is satisfied when σ is sufficiently small. Again, let $A(a) = \gamma a$ for some $\gamma > 1$. Then, the stronger version of (4), condition (17), is also satisfied. By (23), we find

$$a^* = \left[\frac{\gamma(1-\alpha)}{\beta}\right]^{\frac{1-\alpha}{\alpha+\beta-1}}, \qquad \bar{s} = \left[\frac{\gamma(1-\alpha)}{\beta}\right]^{\frac{\beta}{\alpha+\beta-1}}.$$

CONCLUSION

For the standard agency model, this paper proposes an alternative solution to the classical solution in Holmström (1979). There are several advantages of our solution. First, it is a simple closed-form solution and the terms of the solution have straightforward economic interpretations. Second, it is not based on the FOA. This avoids the restrictions implied by the validity of the FOA. Given a set of specific functions, the restrictions from the FOA typically have complex and unclear implications. Finally, it is the first best. The disadvantage of our solution is that it imposes a restriction on the distribution function, although we argue that the requirement of a minimum level of performance for a bonus is consistent with many realistic situations.[2]

The classical solution can be a simple contract, a linear contract, only in a very special case when both the agent and the principal are risk neutral. However, it is shown by Kim–Wang (1998, 2004) that, if there is tiny risk aversion, the optimal contract is completely different from this linear contract. In contrast, our bonus contract is robust in the sense that when risk aversion diminishes, its limit is still a bonus contract.

Interestingly, our solution may offer a different view on the recent debate on manager's pay. Economists have generally assumed that to induce the right incentives, a manager's pay must be closely associated with performance. This view is consistent with a FOA-based solution. However, our solution seems to be in favor of the old way (or the German-Japanese way) in which a bonus is associated with some minimum performance and a close association of pay to performance is unnecessary.

Our solution looks very much like a typical employment contract, in which a fixed wage rate is given based on a minimum level of education and working experience. We can interpret the boundary term $A(a)$ as the requirement for firm-specific skills. A product may be produced by many different technologies/techniques. The best or the most suitable technology for a firm's product tends to produce the product with the highest quality or the most desirable features and thus yields the highest value in the market. In other words, the value of those units of the product produced by the best technology will surpass a threshold of value dependent on a particular technology. Hence, firm-specific skills can be defined by a minimum output value. To surpass this value, an employee needs some firm-specific skills. To acquire these skills, an employee may need some talent and working experience and may need to invest in education and training programs. Thus, by this viewpoint, in the agency

---

[2] Discussions with similar approaches existed in as early as 1970s. See, for example, Gjesdal (1976) and Gravelle–Rees (1992, p.702). However, to our knowledge, no publication exists for a formal treatment of the standard agency model using a similar approach as ours. This is surprising considering the usefulness of a simple contractual solution.

model, the lower boundary of the domain of the distribution function of output should generally be dependent on the agent's effort. The classical solution in Holmström (1979) cannot allow such dependence.

Finally, our solution may explain franchising, by which a local franchisee pays the franchiser a fee for its brand name, where a brand name is typically associated with a minimum level of quality and market acceptance.

REFERENCES

[1] Bhattacharyya, S., and Lafontaine, F. (1995). Double-Sided Moral Hazard and the Nature of Share Contracts. *RAND Journal of Economics*, 26 (4), 761–781.

[2] Gjesdal, F. (1976). Accounting in Agencies. Unpublished mimeo, Stanford University.

[3] Gravelle, H., and Rees, R. (1992). *Microeconomics*, 2nd, New Yong: Longman.

[4] Grossman, S.J., and Hart, O.D. (1983). An Analysis of the Principal-Agent Problem. *Econometrica,* 51, 7–45.

[5] Holmström, B. (1979). Moral Hazard and Observability. *Bell Journal of Economics,* 10, 74–91.

[6] Jewitt, I. (1988). Justifying the First Order Approach to Principal-Agent Problems. *Econometrica,* 56, 1177–1190.

[7] Kim, S.K., and Wang, S. (1998). Linear Contracts and the Double Moral-Hazard. *Journal of Economic Theory,* 82, 342–378.

[8] Kim, S.K., and Wang, S. (2004). Robustness of a Fixed-Rent Contract in a Standard Agency Model. *Economic Theory*, 24 (1), 111–128.

[9] Mirrlees, J.A. (1999). The Theory of Moral Hazard and Unobservable Behavior — Part I. *Review of Economic Studies*, 66, 3–21, written in 1974.

[10] Rogerson, W.P. (1985). The First Order Approach to Principal-Agent Problem. *Econometrica,* 53, 1357–1368.

[11] Shavell, S. (1979). Risk Sharing and Incentives in the Principal and Agent Relationship. *Bell Journal of Economics,* 10, 55–73.

[12] Sinclair-Desgagne, B. (1994). The First-Order Approach to Multi-Signal Principal-Agent Problems. *Econometrica,* 62, 459–465.

INDEX

C

D

E

F

G

H

I

N

O

P

R

T

U

V

W

Y